Bultaco Competition Bikes Owners Workshop Manual

by Jeff Clew
Member of the Guild of Motoring Writers

Models covered:

Alpina	244 cc and 348 cc
Frontera	244 cc and 363 cc
Pursang	196 cc, 244 cc and 363 cc (Mark 5 onwards)
Sherpa T	244 cc and 326 cc

All the above models from 1972 to 1975

ISBN 978 0 85696 219 6

© J H Haynes & Co. Ltd. 1990

All rights reserved. No part of this book may be reproduced or transmitted in any form or by any means, electronic or mechanical, including photocopying, recording or by any information storage or retrieval system, without permission in writing from the copyright holder.

(219-7P5)

J H Haynes & Co. Ltd.
Haynes North America, Inc

www.haynes.com

Acknowledgements

We are greatly indebted to Sammy Miller for the technical assistance given when this manual was being prepared and for the loan of the Bultaco Sherpa T model used in the photographic sequences. We are also grateful to Bultaco International, of Virginia Beach, USA for the complete set of technical publications they so kindly provided. Martin Penny gave the necessary assistance with the overhaul, using a number of the service tools designed and manufactured by Sammy Miller Equipment, of Gore Road, New Milton, Hants. Although not an essential requirement in every instance, they made many of the dismantling and reassembly tasks much easier. Les Brazier took the photographs and Tim Parker edited the text.

We also wish to acknowledge the help of the Avon Rubber Company, who supplied illustrations and advice about tyre fitting, and Amal Limited, for their carburettor illustrations.

About this manual

The author of this manual has the conviction that the only way in which a meaningful and easy-to-follow text can be written is to carry out the work himself, under conditions similar to those found in the average household. As a result, the hands seen in the photographs are those of the author. Even the machines are not new; examples which have covered a considerable mileage are selected, so that the conditions encountered would be typical of those encountered by the average rider/owner. Unless specially mentioned, and therefore considered essential, Bultaco service tools have not been used. There are invariably alternative means of slackening or removing some vital component when service tools are not available, but risk of damage is to be avoided at all costs.

Each of the five Chapters is divided into numbered Sections. Within the Sections are numbered paragraphs. Cross-reference throughout the manual is quite straightforward and logical. For example, when reference is made 'See Section 5.2' it means Section 5, paragraph 2 in the same Chapter. If another Chapter were meant, the reference would read 'See Chapter 2, Section 5.2'. All photographs are captioned with a section/paragraph number to which they refer, and are always relevant to the chapter text adjacent.

Figure numbers (usually line illustrations) appear in numerical order, within a given Chapter. Fig. 1.1 therefore refers to the first figure in Chapter 1. Left-hand and right-hand descriptions of the machines and their component parts refer to the left and right when the rider is seated, facing forward.

Motorcycle manufacturers continually make changes to specifications and recommendations, and these, when notified, are incorporated into our manuals at the earliest opportunity.

Whilst every care is taken to ensure that the information in this manual is correct no liability can be accepted by the authors or publishers for loss, damage or injury caused by any errors in or omissions from the information given.

Contents

Chapter	Section	Page
Introductory pages	Acknowledgements	2
	About this manual	2
	Ordering spare parts	4
	Introduction to the Bultaco	4
	Routine and competition maintenance	6
Chapter 1/Engine, clutch & gearbox	Specifications	9
	Engine/gearbox removal	11
	Dismantling	12
	Examination & renovation	27
	Reassembly	30
	Fault diagnosis	38
Chapter 2/Fuel system & lubrication	Specifications	40
	Carburettor	42
	Exhaust system	47
	Fault diagnosis	48
Chapter 3/Ignition & lighting systems	Specifications	49
	Contact breakers	50
	Ignition timing	51
	Spark plug	55
	Fault diagnosis	58
Chapter 4/Frame & forks	Front forks	59
	Frame	68
	Swinging arm	69
	Rear suspension units	71
	Fault diagnosis	75
Chapter 5/Wheels, brakes & tyres	Specifications	76
	Brakes	77, 79
	Wheels	79
	Tyres	84
	Fault diagnosis	86
Metric conversion tables		87
Safety first!		90
English / American terminology		91
Index		92

Introduction
to the Bultaco range of on/off road machines

The Bultaco motor cycle company was founded by Senor Francisco Bulto on June 3rd 1958 at his private estate near Barcelona, Spain. Like the nucleus of his employees, Senor Bulto had broken away from Montesa when this latter company had decided to discontinue their road racing activities. By forming his own company, Senor Bulto could dictate his own policies and continue to be actively associated with all aspects of motor cycle sport in which he had a very great interest. Time has proved that he made the right decision, for today the Bultaco name is one of the most popular in trials and motocross events held all over the world.

The Bultaco name first came into real prominence when two riders on 175cc Matador models won a gold metal each in the 1962 International Six Days Trial at Garmisch Partenkirchen. This was only a start. At the end of the 1974 season, the company had won no less than 31 gold metals, 17 silver medals and 26 bronze medals in this and successive annual events. British trials ace Sammy Miller joined Bultaco during 1965 to develop their trials model and made devastating impact by winning all manner of events, including an outright win of the Scottish Six Days Trial in 1965, a win that was repeated in 1967 and 1968.

If anything, the Bultaco is even more popular today and is virtually an essential requirement for the expert rider in both trials and motocross events who wishes to maintain his position at the top of the table. Bultaco has come a long way since Senor Bulto decided to manufacture the machines that carry his name, for this must surely be the greatest story of success in modern times. And it looks like continuing, judging from the results of many sporting events held each weekend.

Ordering spare parts

When ordering parts for any Bultaco model, it is advisable to deal direct with a Bultaco agent, who will be better placed to supply the parts ex-stock and will have the technical experience that may not be available with other suppliers. When ordering parts, always quote the engine and frame numbers in full, since these are essential to identify the model and its date of manufacture. Do not omit any suffixes or prefixes, which will make this coding complete. If possible, keep the broken or worn part that has to be replaced, as it is sometimes needed as a pattern, to help identification.

Always fit parts of genuine Bultaco manufacture, never pattern parts that may seem to have a price advantage. Although pattern parts often appear similar, they frequently give inferior service and may prove more expensive in the long run. Some Bultaco agents may be able to offer certain parts on a service exchange basis, cutting costs and getting the bike on the road with a minimum of delay. This service is available only if the worn parts handed in are suitable for reconditioning.

Some of the more expendable parts such as spark plugs, bulbs, tyres, oils and greases etc, can be obtained from accessory shops and motor factors who have convenient opening hours, charge lower prices, and can often be found not far from home. It is also possible to obtain parts on a mail order basis from a number of specialists who advertise regularly in the motorcycle newspapers and magazines.

In the UK, one of the leading stockists of parts for Bultaco models is Sammy Miller Equipment, of Gore Road, New Milton, Hants, from whom a fully illustrated catalogue is available. The list includes a number of service tools which, whilst not essential, will make the dismantling and reassembly sequence much easier. If a machine is worked on regularly, they will represent a wise investment.

Frame number

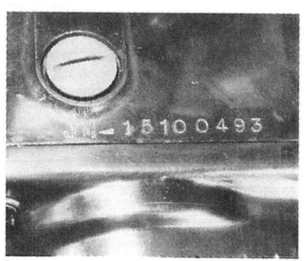

Engine number

Left- and right-hand views of the 1975 250 cc Bultaco Sherpa T

Routine maintenance

Periodic routine maintenance is a continuous process that commences immediately the machine is used. It must be carried out at specified mileage recordings or on a calendar date basis if the machine is not used regularly, at whichever falls soonest. Maintenance should be regarded as an insurance policy, to help keep the machine in peak condition and to ensure long, trouble-free service. It has the additional benefit of giving early warning of any faults that may develop and will act as a regular safety check, to the obvious advantage of both rider and machine alike.

Because the models covered by this manual are designed mainly for competition or general off-road use, it is not practicable to list the various maintenance tasks related to the mileage covered, except in a few isolated cases. For this reason, a different format has been drawn up, in the knowledge that the machine will have to be prepared just prior to a competition event and then checked during the various phases of the event itself, in some cases. Different models are used for different types of event and so the maintenance tasks have been arranged under the respective model headings. It should be remembered that the interval between the various maintenance tasks serves only as a guide. As the machine gets older or is used under particularly adverse conditions, it would be advisable to reduce the period between each check.

Some of the tasks are described in detail where they are not mentioned fully as a routine maintenance item in the text. If a specific item is mentioned, but not described in detail, it will be covered fully in the appropriate chapter. No special tools are required for the normal routine maintenance tasks. The tools contained in the tool kit supplied with every new machine will prove adequate for each task or if they are not available, the tools found in the average household will usually suffice.

Before and after each event

Alpina models

Each of the following parts should be lubricated, using oil of the viscosity recommended in each case:

Rear chain - use an aerosol-type chain lubricant. If the chain is very dirty, remove it, clean it and then immerse it in a bath of molten lubricant to achieve real penetration.
Handlebar levers - lubricate joints and cable nipples with SAE 10 oil.
Control cables - lubricate with SAE 10 oil. If necessary, remove cables and suspend them.
Primary chaincase - drain off old oil and refill with 300ccs SAE 30 oil.
Gearbox - drain off old oil and refill with 500ccs SAE 90 gearbox oil.
Swinging arm fork - grease until clean grease exudes from the joints.
Magneto felt pad - remove the magneto cover and lubricate the felt pad with a few drops of distributor oil.
Twistgrip - remove and grease the assembly thoroughly.
Front brake - remove the brake plate from the wheel, clean and grease the operating pivot sparkingly. Pack the inner boss of the brake plate with grease, prior to reassembly.
Rear brake - as front brake.
Speedometer cable - remove, withdraw the inner cable and grease the latter. Do not grease within the six inches of the point where the cable enters the speedometer head.

After six events

Complete all the previous tasks, then grease and adjust the steering head bearings, dismantle and pack the wheel bearings with fresh grease and drain and refill the front forks with 175ccs SAE 30 oil per fork leg.

Every 500 miles

Frontera models

Each of the following parts should be lubricated, using oil of the viscosity recommended in each case:
Rear chain - use an aerosol-type chain lubricant. If the chain is very dirty, remove it, clean it and then immerse it in a bath of molten lubricant to achieve real penetration.
Handlebar levers - lubricate joints and cable nipples with SAE 10 oil.

Every 1,500 miles

Complete all the previous tasks, then grease the twistgrip assembly thoroughly after it has been dismantled. Remove and withdraw the inner cable of the speedometer drive and grease it. Do not grease within six inches of the point where the cable enters the speedometer head. Detach the magneto cover and lubricate the felt pad with a few drops of distributor oil. Lubricate all the control cables with SAE 10 oil, removing them if necessary so that the oil can penetrate right through when they are suspended.

Every 3,000 miles

Complete the 500 mile and 1,500 mile checks, then drain the primary chaincase and refill it with 300ccs of SAE 30 oil. Drain the gearbox and refill it with 500ccs of SAE 90 gearbox oil. Remove the front and rear brake plates and grease the brake operating pivot sparingly. Pack the inner boss of the brake place with grease, prior to reassembly.

Unscrew this cover to adjust clutch

Slacken locknut, adjust, then tighten and re-check

Filling the primary chaincase

The gearbox filler plug

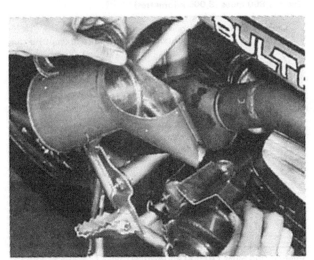
Home made funnel makes gearbox filling easier

Every 8,000 miles

Complete the 500 mile, 1,500 mile and 3,000 mile tasks, then drain and refill the front forks with 175ccs of SAE 30 oil per fork leg. Dismantle the wheel bearings, repack them with fresh grease and replace them. Grease and adjust the steering head bearings.

Maintenance before and after each heat

Pursang models
Rear chain - lubricate the rear chain with an aerosol-type lubricant or alternatively, from an oil can of the pressure feed type. Apply the lubricant to the top of the lower chain run, immediately in front of the sprocket, whilst the wheel is slowly rotated by hand.
Handlebar levers - lubricate joints and cable nipples with SAE 10 oil. Run some oil down each cable inner.
Twistgrip - dismantle the twistgrip assembly and lubricate with SAE 10 oil, allowing some to run down the inner cable.

Routine Maintenance

Maintenance before and after each race

Twistgrip - dismantle the twistgrip assembly, grease it thoroughly and then reassemble it, after checking the condition of the control cable.
Primary chaincase - drain the primary chaincase and refill it with 300ccs of SAE 30 oil.
Gearbox - drain the gearbox and refill it with 600 ccs of SAE 90 gearbox oil.
Front brake - remove the brake pedal from the wheel, clean and grease the operating pivots, then pack the inner boss of the brake plate with grease, prior to assembly.
Rear brake - as front brake.

Maintenance after 6 races

Steering head bearings - grease the head race bearings and adjust, if necessary.
Wheel bearings - dismantle the hubs and pack the wheel bearings with fresh grease, then reassemble.
Front forks - drain the forks and refill each fork leg with the correct quantity of fork oil, refer to page 59 for oil capacity and grade.

Before and after each round

Sherpa models
Rear chain - use an aerosol-type chain lubricant or alternatively an oil can of the pressure feed type, to lubricate the top of the lower chain run whilst the wheel is slowly rotated by hand. If the chain is very dirty, remove and clean it, then immerse it in molten lubricant, to ensure thorough penetration.
Handlebar levers - lubricate the joints and cable nipples with SAE 10 oil. In the case of the twistgrip, allow some oil to run down the cable, after dismantling the twistgrip assembly. Lubricate the twistgrip at the same time.

Before and after each complete trial

Primary chaincase - drain and refill with 300ccs SAE 5 or 10 oil.
Gearbox - drain and refill with 500ccs SAE 90 gearbox oil.
Magneto felt pad - remove the magneto cover and lubricate the felt pad with a few drops of distributor oil.
Twistgrip - dismantle the twistgrip assembly and grease it thoroughly.
Front brake - remove the brake plate from the wheel, clean and grease the operating pivot sparingly. Pack the inner boss of the brake plate with grease, prior to reassembly.
Rear brake - as front brake.
Speedometer cable - remove, withdraw the inner cable and grease the latter. Do not grease within six inches of the point where the cable enters the speedometer head.
Control cables - oil all the inner cables with SAE 10 oil. If necessary, remove the cables and suspend them, so that the oil will penetrate throughout.
Tyres - If the leading edges of the tread have worn, reverse the tyres on the wheel rims so that the worn area is now trailing.

After 6 trials

Complete all the maintenance tasks carried out after each trial, then:

Steering head bearings - grease the head race bearings and adjust if necessary.
Wheel bearings - dismantle the hubs and pack the wheel bearings with fresh grease, then reassemble.
Front forks - drain the forks and refill each fork leg with the correct quantity and grade of fork oil.
Note: It follows that in addition to the tasks listed, the machine should always be examined thoroughly and any loose nuts, bolts or screws tightened. Pay particular attention to the wheel spokes, the accuracy of the ignition timing and the cleanliness of the carburettor. The flywheel securing nut is also prone to work loose unless tightened to the correct torque setting and even then a regular check is necessary. Do not overlook the statutory requirements either, if the machine is to be used on the road. The lights, if fitted, must be in full working order and there must also be a horn that will give audible warning of approach. The machine must be adequately silenced and the tyres must conform to the regulations currently in force, especially with regard to the 2 mm minimum tread depth.

General maintenance guide

Where it is practicable to base the routine maintenance schedules on a mileage basis, the following guidelines may help. Some of the items included have not been listed previously, mainly because it is difficult to recommend the frequency with which these tasks are necessary.

Every 600 miles (1,000 kilometres)

Clean and re-set the spark plug(s).
Clean the carburettor air cleaner.
Clean the carburettor feed pipe filter.
Re-tension and lubricate the final drive chain.

Every 1,800 miles (3,000 kilometres)

Complete all the 600 mile tasks, then:
Clean and re-gap the contact breaker points (except Pursang models).
Clean the petrol tap filter(s).

Every 3,000 miles (5,000 kilometres)

Complete all the 600 mile and 1,800 mile maintenance tasks, then:
Decarbonise the engine and exhaust system.
Drain and refill the primary chaincase.
Drain and refill the gearbox.

Every 6,000 miles (10,000 kilometres)

complete the 600 mile, 1,800 mile and 3,000 mile maintenance tasks, then:
Clean out the petrol tank.
Check and if necessary adjust the steering head bearings.
Change all of the control cables.

Chapter 1 Engine, clutch and gearbox

Contents

General description ... 1	Examination and renovation: crankcase castings ... 23
Operations with engine in frame ... 2	Examination and renovation: gearbox components ... 24
Operations with engine removed ... 3	Examination and renovation: clutch actuating mechanism ... 25
Method of engine/gearbox removal ... 4	Examination and renovation: clutch assembly ... 26
Removing the engine/gear unit ... 5	Examination and renovation: primary chain ... 27
Dismantling the engine/gear unit: general ... 6	Examination and renovation: gear selector mechanism ... 28
Dismantling the engine/gear unit: removing the cylinder head, barrel and piston ... 7	Reassembly: general ... 29
Dismantling the engine/gear unit: removing the flywheel magneto ... 8	Reassembling the engine/gear unit: replacing the kickstart shaft and spring ... 30
Dismantling the engine/gear unit: removing the gearbox final drive sprocket ... 9	Reassembling the engine/gear unit: replacing the crankshaft assembly ... 31
Dismantling the engine/gear unit: removing the kickstart return spring ... 10	Reassembling the engine/gear unit: replacing the gear cluster ... 32
Dismantling the engine/gear unit: dismantling the clutch ... 11	Reassembling the engine/gear unit: joining the crankcase halves ... 33
Dismantling the engine/gear unit: separating the crankcase ... 12	Reassembling the engine/gear unit: replacing the gear selector pawls ... 34
Dismantling the engine/gear unit: removing the crankshaft assembly ... 13	Reassembling the engine/gear unit: replacing the clutch and primary drive ... 35
Dismantling the engine/gear unit: removing the gearbox components ... 14	Reassembling the engine/gear unit: replacing the flywheel magneto ... 36
Dismantling the engine/gear unit: removing the crankshaft and gearbox main bearings ... 15	Reassembling the engine/gear unit: replacing the final drive sprocket ... 37
Examination and renovation: general ... 16	Reassembling the engine/gear unit: replacing the piston, cylinder barrel and cylinder head ... 38
Examination and renovation: main bearings and oil seals ... 17	Replacing the engine/gear unit in the frame ... 39
Examination and renovation: crankshaft assembly ... 18	Starting and running the rebuilt engine ... 40
Examination and renovation: cylinder barrel ... 19	Fault diagnosis: engine ... 41
Examination and renovation: piston and piston rings ... 20	Fault diagnosis: clutch ... 42
Examination and renovation: small end bearing ... 21	Fault diagnosis: gearbox ... 43
Examination and renovation: cylinder head ... 22	

Specifications

The Bultaco models covered in this manual (Alpina, Frontera, Pursang and Sherpa) employ the same basic engine/gearbox unit in which the gearbox is an integral part of the engine assembly. Although the information given relates specifically to the 350 cc Sherpa model used in the photographic sequences, it applies equally well to the other models, even though the engine unit may have different bore and stroke dimensions and the gearbox, different gear ratios.

A similar dismantling and reassembly procedure is applicable to all the models covered. Where any significant changes in design have been made, mention is included in the text, together with the revised procedure necessary.

The Chapter concludes with a fault finding table.

Engine

	Alpina	Frontera	Pursang	Sherpa
Model				
Capacity ccs	244 (250)	244 (250)	196 (200)	244 (250)
	348 (350)	363 (360)	244 (250)	326 (350)
			363 (360)	
Bore and stroke mm				
200 cc	—	—	64.5 x 60	—
250 cc	72 x 60	72 x 60	72 x 60	72 x 60
350 cc	83.2 x 64	—	—	83.2 x 60

	Alpina	Frontera	Pursang	Sherpa
360 cc	—	85 x 64	85 x 64	—
Compression ratio				
200 cc	—	—	13 : 1	—
250 cc	9 : 1	12 : 1	12 : 1	9 : 1
350 cc	9.5 : 1	—	—	9 : 1
360 cc	—	10.5 : 1	11 : 1	—
BHP @ rpm				
200 cc	—	—	24.6 @ 7,000	—
250 cc	19.8 @ 5,500	31.2 @ 8,000	34.3 @ 7,000	20 @ 5,500
350 cc	22.7 @ 6,000	—	—	21.5 @ 5,500
360 cc	—	33.5 @ 8,000	40 @ 7,000	—

Gear ratios

Alpina 250cc and 350cc
- 1st 0.287 : 1
- 2nd 0.422 : 1
- 3rd 0.625 : 1
- 4th 0.882 : 1
- 5th 1 : 1

Frontera 250cc and 360cc
- 1st 0.287 : 1
- 2nd 0.422 : 1
- 3rd 0.625 : 1
- 4th 0.821 : 1
- 5th 1 : 1

Pursang 200cc, 250cc and 360cc
- 1st 0.376 : 1
- 2nd 0.513 : 1
- 3rd 0.670 : 1
- 4th 0.832 : 1
- 5th 1 : 1

Sherpa 250cc and 350cc
- 1st 0.263 : 1
- 2nd 0.342 : 1
- 3rd 0.442 : 1
- 4th 0.723 : 1
- 5th 1 : 1

Sprocket sizes

	Alpina		Frontera		Pursang			Sherpa	
Model	250	350	250	360	200	250	360	250	350
Engine sprocket	16	16	16	16	16	16	16	16	16
Clutch sprocket	38	38	38	38	38	38	38	38	38
Gearbox sprocket	12	13	12	13	11	12	13	11	11
Rear wheel sprocket	42	42	42	42	46	46	46	46	46

Gearbox oil content ... 600 ccs SAE 90
Primary drive case ... 300 ccs SAE 30 (SAE 5 or 10, Sherpa models only)
Chain (primary)
- Make ... Joresa or Regina
- Type ... Duplex, 3/8 inch pitch
- No. of links ... 52

1 General description

The engine/gearbox unit fitted to the Bultaco models is of the single cylinder two-stroke type, using a flat top piston, hemispherical cylinder head and what is known as loop scavenging to effect a satisfactory induction and exhaust sequence. The shape and arrangement of the cylinder ports guarantees a very high standard of performance without the need for mechanical aids such as rotary or reed induction valves. The aluminium alloy piston is fitted with two compression rings, each pegged in characteristic two-stroke fashion to prevent rotation and risk of the ends being trapped in the cylinder ports.

The light alloy cylinder barrel is fitted with a cast iron sleeve and aluminium alloy is used extensively for all the engine and gearbox castings, giving efficient cooling and lightness in weight. The connecting rod small end and big end bearings are of the caged roller type; journal ball bearings are used for the crankshaft main bearings, the 1973/4 models have two supporting the drive side and one on the flywheel magneto side of the rigid, built-up crankshaft. The flywheel magneto is located on the left-hand side of the machine and is of the ac generator type, the flywheel rotor being attached to the end of the crankshaft.

Lubrication is effected by the petroil system, a simple but efficient system in which a measured amount of self-mixing oil is dissolved in the petrol. Because a two-stroke engine relies on crankcase compression as part of the cycle of operation, the oil content of the petrol is distributed to the working parts of the engine. The system works on the total loss principle, all excess oil being discharged via the exhaust. The crankcase is effectively

Chapter 1: Engine, clutch and gearbox

sealed by oil seals fitted to each side of the crankshaft.

The routing of the exhaust system contrasts sharply. On some machines, usually those of the trials or enduro type, it is swept over the top of the cylinder head and carried on the right-hand side of the machine, neatly tucked within the frame. Models of the motocross or cross country type have a downswept system, which passes below the crankcase of the engine unit and emerges on the right-hand side. Either one or two silencers are included in the specification. Primary transmission is by chain running from a sprocket on the right-hand side of the crankshaft to the clutch assembly on the end of the gearbox mainshaft. Because the engine is built on the unit construction principle, the gearbox cannot be moved independently of the engine to make adjustments in chain tension. An automatic tensioner is provided to keep the tension correct during service; when chain wear develops, renewal is essential.

The multi-plate clutch assembly runs in oil, driving a five-speed constant mesh gear train which runs on ball journal bearings. The clutch assembly comprises twelve all metal discs, six driving and six driven. A kickstart lever is fitted to the left-hand side of the machine and a gear change lever to the right-hand side, the latter assembly incorporating a positive stop mechanism to ensure each gear is selected in sequence, with positive action. Separate oil contents are provided for the primary chaincase and the gearbox, each holding a small quantity of oil that must be changed at prescribed intervals.

2 Operations with engine in frame

It is not necessary to remove the engine unit from the frame unless the crankshaft assembly and/or the gearbox components require attention. Most operations can be accomplished with the engine in place, such as:

1. *Removal and replacement of cylinder head.*
2. *Removal and replacement of cylinder barrel and piston.*
3. *Removal and replacement of clutch assembly.*
4. *Removal and replacement of flywheel generator.*
5. *Removal and replacement of contact breaker assembly.*
6. *Removal and replacement of gear selector shaft, return spring and selector pawl assembly, also the kickstart return spring.*

When several operations need to be undertaken simultaneously, it will probably be advantageous to remove the complete engine/gear unit from the frame. This will provide the dual advantage of better access and more working space.

3 Operations with engine removed

1. Removal and replacement of the main bearings.
2. Removal and replacement of the crankshaft assembly.
3. Removal and replacement of the gear cluster, selectors and gearbox main bearings.

4 Method of engine/gearbox removal

As described previously, the engine and gearbox are built in unit and it is necessary to remove the unit complete, in order to gain access to either component. Separation is accomplished after the engine has been removed and refitting cannot take place until the crankcases have been reassembled. Note that when the crankcases are separated, the gearbox internals will be exposed. The crankcase assembly cannot be dismantled completely without stripping the gearbox, and vice-versa.

5 Removing the engine/gearbox unit

1 As no centre stand is provided except on the Frontera model additional means will be necessary to secure the machine

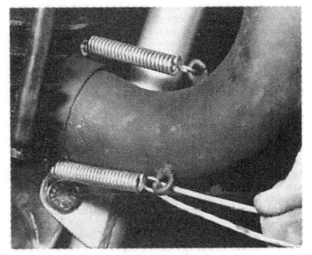
5.6 Use a piece of stout wire to unhook springs

5.8 Remove carburettor as a complete unit initially

5.10 Kickstart is retained by a pinch bolt

by placing wooden blocks under the frame so that both wheels are off the ground. Alternatively a stand can be fabricated for this purpose, which will come in useful on future occasions. The machine must stand firmly when unsupported other than by the stand, and must be on level ground.

2 Drain the oil from the primary chaincase and the gearbox, preferably while it is warm. The latter has a drain plug and washer under the left-hand crankcase and the former a drain plug and washer under the clutch assembly. Before removing either plug, it is advisable to clean the surrounding area, to prevent the ingress of dirt.

3 Turn off the fuel supply and loosen the clip that retains the nylon fuel pipe to the base of the fuel tap (two taps and two fuel pipes, Frontera and Pursang models). Remove the fuel pipe(s). At the front of the tank, loosen the hose clip securing the nylon pipe that interconnects both halves of the fuel tank. Pull off the pipe and allow the fuel to drain into a suitable receptacle. Alternatively, drain the fuel tank completely before the pipe is detached.

4 Remove a screw, washer, rubber washer and nut from each side of the seat fairing and remove the nut, washer and rubber washer from the top front end of the tank, which retains the tank to the top frame tube. Detach the vent pipe from the top of the tank, then carefully lift off the complete tank and seat assembly. Store it in a safe place, away from any naked lights.

5 Using a piece of stout wire, stretch the single coil spring that secures the forward section of the exhaust system to the rear portion. Remove the coil spring. Take out the bolt, washer and nut that attach the rear silencer to the frame and remove the rear silencer. The Pursang and Frontera models have a different type of exhaust system, the silencer of which is bolted to a stay extending from the top of the right-hand rear suspension unit mounting.

6 The front section of the exhaust system is retained by three coil springs, which should be removed in similar fashion. They retain the exhaust pipe to the cylinder exhaust stub. Remove the nut, washer and bolt that secure the exhaust system to the rear of the gearbox and remove the exhaust system complete. This bolt is one of several attaching the gearbox to the frame.

7 Remove the bolt and washer that secures the air cleaner to the frame. Note that on the Pursang and Frontera models it will be necessary to detach the side panels that contain the riding number - each is held in position by three screws. Remove the two clips and the air intake hose from the carburettor. The air cleaner complete with the anti-chafe seat rubbers can now be withdrawn.

8 Remove the two nuts and washers from the carburettor flange, or if the carburettor is of the stub fitting type, slacken the screw in each of the hose clips. Draw off the carburettor as a complete unit and permit it to hang by means of the throttle cable. It is advisable to loop the cable over the frame and perhaps tie the carburettor in position, so that it is kept well clear of the engine.

9 Remove the two bolts and washers that secure the front portion of the rear chainguard to the flywheel magneto cover and the frame. Remove the chainguard then the rear chain, preferably by positioning the spring link on the rear wheel sprocket so that removal of the link is much easier.

10 Slacken the pinch bolts that retain the kickstart and gearchange levers, after first marking their positions to aid reassembly. The kickstarter bolt has a nut and there is a felt washer behind the gearchange lever. Pull each lever off its splined shaft.

11 Disconnect the clutch cable from the handlebar lever, which will provide sufficient slack for the cable to be removed from the clutch operating arm on the gearbox. Tie the cable so that it is clear of the engine unit. Remove the small chainguard over the gearbox sprocket.

12 Detach the lead from the spark plug and disconnect the four leads from the flywheel magneto at the terminal block found on the front down tube. Models fitted with electric lighting may have additional leads, but each is colour-coded to aid eventual replacement in the correct order.

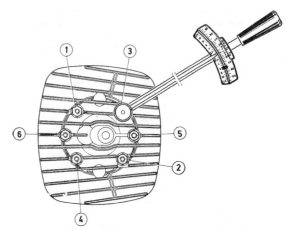

Fig. 1.1 Sequence for loosening and tightening the cylinder head nuts and bolts

13 Loosen the cylinder head securing nuts and bolts one half-turn only, following the sequence shown in the accompanying illustration. Repeat this procedure three times until each nut or bolt has loosened by two complete turns. Remove the cylinder head steady by withdrawing nut 2 and washer and bolt 5 and washer. Mark each with a scriber to ensure they are replaced in the identical positions when reassembly eventually takes place.

14 The engine unit is now retained to the frame by either three or four bolts, depending on the model, in front and behind the engine unit. One of these bolts may already have been removed when the exhaust system was taken off. Remove the remaining nuts, washers and bolts, noting that the engine will drop slightly when the front bolt(s) is removed. Lift the engine vertically approximately two inches, tilt it slightly until the front engine mounting lug is just clear of the frame lug, then lift the engine out of the frame from the right-hand side.

6 Dismantling the engine/gear unit - general

1 Before commencing work on the engine unit, the external surfaces should be cleaned thoroughly. A motor cycle engine has very little protection from road grit and other foreign matter, which will find its way into the dismantled engine if this simple precaution is not observed. One of the proprietary cleaning compounds such as 'Gunk' can be used to good effect, particularly if the compound is allowed to work into the film of oil and grease before it is washed away. When washing down, make sure that water cannot enter the carburettor or the electrical system, particularly if these parts have been exposed.

2 Never use undue force to remove any stubborn part, unless mention is made of this requirement. There is invariably good reason why a part is difficult to remove, often because the dismantling operation has been tackled in the wrong sequence. Dismantling will be made easier if a simple engine stand is constructed that will correspond with the engine mounting points. This arrangement will permit the complete unit to be clamped rigidly to the work bench, leaving both hands free.

7 Dismantling the engine/gear unit: removing the cylinder head, barrel and piston

1 Remove the remaining nuts, bolts and washers from the cylinder head and lift the cylinder head off its retaining studs. Identify the nuts and bolts with their respective numbers, so that they are replaced in an identical location, during reassembly.

2 Tap the cylinder barrel very gently with a raw hide mallet in the vicinity of the exhaust stub, in order to break the joint between the base of the cylinder barrel and the crankcase. Carefully lift the cylinder away from the crankcase, taking the pre-

5.10a Gearchange lever has similar means of attachment

5.11 Small chainguard covers the gearbox sprocket

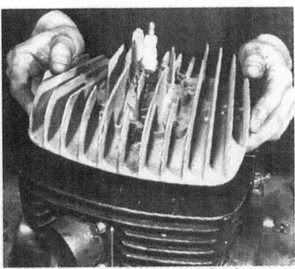
7.1 Cylinder head nuts and bolts must be slackened in sequence

7.2 Catch piston as it emerges from the bore

7.3 Use long nose pliers to remove circlip

7.3a Small end is of the caged needle roller type

14

Chapter 1: Engine, clutch and gearbox

caution of covering the mouth of the crankcase with clean rag during the initial lift so that nothing can drop into the crankcase - an important consideration if the piston rings have broken. Catch the piston as it emerges from the bore, or there is risk of damage to the piston and/or the piston rings.

3 Remove the piston complete with rings by detaching one of the gudgeon pin circlips with a pair of long nose pliers. Tap the gudgeon pin out of position with a soft metal drift, whilst supporting the piston. If the gudgeon pin tends to be a tight fit, warm the crown of the piston first with a rag soaked in hot water and wrung out. This should expand the piston bosses sufficiently to release their grip on the gudgeon pin. When removing the piston from the connecting rod, collect the thrust washers fitted to each end of the small end, otherwise they will drop into the crankcase mouth.

8 Dismantling the engine/gear unit: removing the flywheel magneto

1 Remove the four Allen screws and their washers from the

Fig. 1.2 Crankshaft, cylinder barrel and cylinder head

1 Crankshaft assembly complete
2 Magneto side crank
3 Clutch side crank
4 Crankpin
5 Thrust washer - 2 off
6 Crankpin end plug - 2 off
7 Piston ring - 2 off
8 Big-end rollers - 12 or 13 off
9 Cylinder head washer - 6 off
10 Small end bearing
11 Engine sprocket
12 Flywheel (inner)
13 Flywheel (outer)
14 Inlet manifold
15 Dowel pin - 2 off
16 Dowel pin - 2 off
17 Dowel pin
18 Cylinder base gasket
19 Cylinder assembly
20 Cylinder liner
21 Inlet manifold - stub type
22 Inlet manifold gasket
23 Cylinder head
24 Piston
25 Left-hand crankcase
26 Right-hand crankcase
27 Piston ring - 2 off
28 Gudgeon pin
29 Circlip - 2 off
30 Oil seal cover
31 Oil seal cover
32 Oil seal cover - drive side
33 Gudgeon thrust washer - 2 off
34 Connecting rod assembly
35 Crankshaft bush
36 Flywheel nut
37 Shouldered flywheel nut
38 Rear engine plate
39 Crankcase gasket
40 Front engine bolt assembly
41 Generator stator plate
42 Head steady assembly
43 Grommet
44 Grommet
45 Drain plug
46 Oil seal cover gasket - 2 off
47 'O' ring
48 Screw - 3 off
49 Screw - 12 off
50 Bolt - 2 off
51 Bolt
52 Nut - 16 off
53 Flat washer - 4 off
54 Flat washer - 3 off
55 Lock washer - 3 off
56 Main bearing - 3 off (2 on some models)
57 Oil seal - Viton
58 Oil seal - Viton
59 Oil seal - Standard type
60 Oil seal - Viton
61 Allen screw
63 Star washer - 5 off
64 Star washer - 24 off
65 Star washer
66 Locknut
67 Countersunk star washer - 6 off
68 Nylon locknut - 4 off
69 Nylon locknut - 2 off
70 Nylon locknut - 2 off
71 Nut - 6 off
72 Flat washer - 3 off
73 Stud - 4 off
74 Stud - 2 off
75 Stud - 2 off
76 Stud - 4 off
77 Stud - 4 off
78 Stud - 3 off
79 Stud - 9 off
80 Woodruff key

Note: The quantity of some of the parts listed may vary according to the model designation

8.1 Magneto cover is retained by four Allen screws

8.3 Special tool provides convenient method of holding flywheel rotor

8.3a Retaining nut has spring washer on underside

8.4 Extractor tool is necessary to avoid damage to rotor

9.1 Final drive sprocket is retained by large nut

9.2 Wrap chain around sprocket to lock whilst removing nut

left-hand crankcase cover, then remove the cover itself. Note that the rearmost Allen screw is the shortest.

2 Remove the felt washer from the end of the clutch push rod at the drive sprocket nut, and the felt washer from the crankcase boss for the rear Allen screw.

3 Hold the flywheel rotor steady, preferably by the use of a special flywheel holding tool as shown in the accompanying photograph. Slacken the centre retaining nut (right-hand thread) and remove it, together with the spring washer and plain washer beneath. Remove the flywheel holding tool.

4 Fit the flywheel puller and turn the puller bolt clockwise to free the flywheel from the tapered end of the crankshaft. Pull the flywheel from the taper and remove the Woodruff key from the crankshaft. Note that it is essential to use a puller to remove the flywheel rotor; other improvised techniques will almost certainly cause damage because the flywheel rotor is a very tight fit.

5 Before removing the stator plate, which is retained by three screws around its periphery, mark its position in relation to the surrounding crankcase. If the marking is clear and accurate, it will obviate the need to retime the ignition on reassembly of the engine.

9 Dismantling the engine/gear unit: removing the gearbox final drive sprocket

1 The gearbox final drive sprocket is secured to the end of the gearbox mainshaft by a large hexagon nut and tab washer. Bend back the tab washer and unscrew the nut, which has a right-hand thread.

2 In order to lock the sprocket firmly during this operation, wrap the rear chain around it as shown in the accompany photograph. When the nut has been removed, the sprocket will pull off the splined mainshaft.

10 Dismantling the engine/gear unit: removing the kickstart return spring

1 Unscrew the locknut and the special flanged retaining nut which holds one end of the return spring against the crankcase.
2 Using a piece of stout wire shaped like a hook, pull the other end of the spring from the kickstart shaft and remove the spring.

11 Dismantling the engine/gear unit: dismantling the clutch

1 Remove the eight Allen screws and washers from the primary

11.1 Remove the eight Allen screws that retain primary drive cover

11.2 Part of gear selector mechanism will remain within cover

11.4 Special tool makes compression of clutch springs easy

11.4a Cutaway permits withdrawal of pin

11.5 Lift off pressure plate ...

11.5a ... followed by individual clutch plates ...

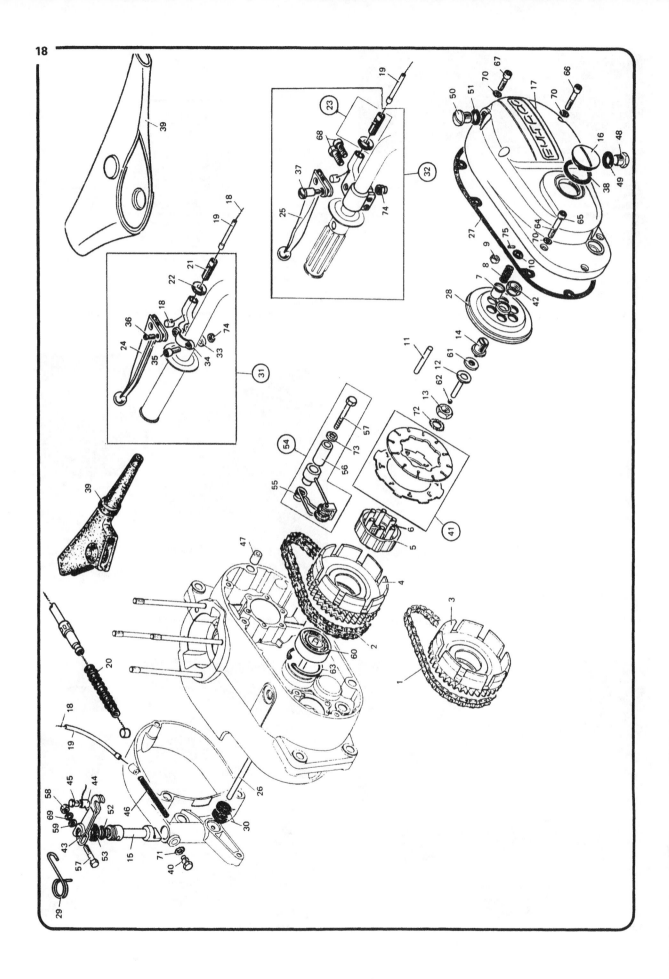

Chapter 1: Engine, clutch and gearbox

drive cover on the right-hand crankcase and remove the cover. The screws are of different lengths and it is therefore advisable to make a note of their arrangement, to aid subsequent re-assembly.

2 Place a container below the cover to catch the oil remaining inside, then break the joint by tapping the cover lightly with a raw hide mallet. Lift the cover away, taking care not to damage the sealing gasket.

3 Remove the bolt and washer so that the chain tensioner assembly and spring can be lifted out.

4 Compress the clutch pressure plate, preferably by the use of a special spring compressor which makes this task very much easier. Compress each spring in turn and withdraw the retaining pin, which is easily lost. Remove the spring cap and spring, then repeat until the remaining five springs have been removed in similar fashion.

5 Remove the pressure plate and detach the bearing from the back of the plate. Withdraw all the clutch plates, twelve in total, then remove the clutch pushrod end piece (mushroom) from the gearbox mainshaft. If the pushrod ball is not retained in the end, press the main pushrod through from the left-hand side of the

Fig. 1.3 Clutch and primary transmission

1 Primary chain - single row type
2 Primary chain - double row type
3 Clutch drum, complete with sprocket - single row type
4 Clutch drum, complete with sprocket - double row type
5 Clutch centre, complete with sprocket
6 Clutch centre stud - 6 off
7 Thimble - 6 off
8 Clutch spring - 6 off
9 Clutch spring nut - 6 off
10 Clutch spring washer - 6 off
11 Clutch pushrod extension
12 Clutch pushrod end piece (mushroom)
13 Clutch centre nut
14 Clutch adjusting screw
15 Clutch operating arm
16 Clutch adjusting screw cap
17 Clutch cover
18 Clutch inner cable
19 Clutch outer cable
20 Nylon cable assembly
21 Clutch cable adjuster
22 Adjuster screw
23 Clutch cable adjuster assembly
24 Metal clutch lever
25 Forged alloy clutch lever
26 Clutch operating rod
27 Clutch cover gasket
28 Clutch pressure plate
29 Clutch operating arm return spring
30 Felt washer
31 Metal clutch lever assembly
32 Forged alloy clutch lever assembly
33 Clutch lever support
34 Clamp
35 Screw - 2 off
36 Pivot bolt - metal assembly
37 Pivot bolt - forged alloy assembly
38 Adjusting screw cap gasket
39 Clutch lever cover
40 Stop bolt
41 Clutch plate set - 6 plates
42 Clutch adjusting nut
43 Clutch operating arm lever
44 Cable nipple
45 Cable nipple screw
46 Clutch cable return spring
47 Dowel pin - 2 off
48 Drain plug
49 Drain plug washer
50 Filler cap
51 Filler cap gasket
52 'O' ring
53 Felt washer
54 Primary chain tensioner assembly
55 Primary chain tensioner spring
56 Primary chain tensioner spacer
57 Bolt
58 Nut
59 Flat washer
60 Gearbox bearing
62 Ball bearing
63 Spring clip
64 Allen screw - 2 off
65 Allen screw
66 Allen screw - 4 off
67 Allen screw
68 Allen screw - 2 off
69 Star washer
70 Star washer - 8 off
71 Star washer
72 Star washer
73 Star washer
74 Nylon locknut
75 Steel pin - 6 off

Note: The quantity of some parts listed may vary according to the model designation

11.5b ... which are of the all-metal type, then ...

11.5c ... lift out pushrod end piece

11.6 Cut locking wire on flywheel nut

11.7 Clutch drum tool will aid removal of centre nut

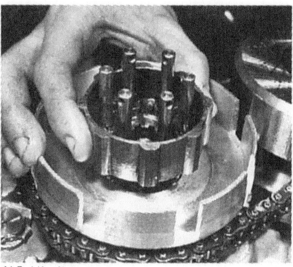
11.7a Lift off the clutch centre

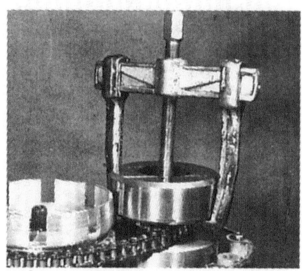
11.8 Puller is needed to draw flywheel off tapered crankshaft end

11.8a Chain is endless, therefore clutch and flywheel must be lifted off in unison

11.9 Six screws and washers retain the oil seal assembly

machine until the ball is displaced from the mainshaft. This too is small and easily lost.

6 Cut the locking wire from the nut that retains the engine sprocket and flywheel assembly in position. Using the clutch holding tool again, remove the nut and washer beneath.

7 With the clutch drum holding tool still in position, slacken and remove the clutch centre retaining nut and star washer. Lift off the clutch centre.

8 Remove the clutch holding tool and pull the clutch housing and drive sprocket/flywheel assembly off their respective shafts together with the primary chain. The two actions must be co-ordinated because the primary chain is of the endless type.

9 Remove the Woodruff key and flanged spacer from the crankshaft. Slacken and remove the six screws and washers that secure the oil seal retainer and lift off the retainer, complete with oil seal and 'O' ring. Remove the second 'O' ring from the crankshaft and the flanged spacer from the gearbox mainshaft.

10 Pull out the gear selector shaft and return spring from the crankcase cover. Remove the three screws from the gear selector pawl cover plate. Withdraw the selector pawl assembly from the crankcase, as shown in the accompanying photograph.

12 Dismantling the engine/gear unit: separating the crankcases

1 Remove the eleven nuts and washers, one Allen screw and washer and one Nyloc nut and washer from the left-hand crankcase. The star washers can be removed with a magnet, if difficulty is experienced.

2 Heat around the right-hand main bearing to approximately 300 - 350°F (150 - 180°C). If using a localised source of heat, such as a blowlamp, keep it moving so that the heat is not concentrated in one area. When the crankcase is hot enough, hold the engine unit right-hand side uppermost and lightly tap the ends of the crankshaft and gearbox mainshaft with a raw hide mallet until the crankcases separate. **No undue force should be necessary.**

Note: Never attempt to lever the crankcases apart with a screwdriver because this will cause irreparable damage to the jointing surfaces. Always remember that the efficiency of a two-stroke engine is dependent on a leak-proof crankcase joint.

3 When the crankcases have separated and cooled down, the crankshaft and main bearings, together with the gear cluster, will remain in the left-hand crankcase. Only the gearbox mainshaft and layshaft bearings will remain in the right-hand crankcase.

11.9b Withdraw flanged spacer from gearbox mainshaft

11.10 Withdraw the selector pawl assembly

11.9a Underside of the oil seal retainer

12.2 Do not use force when separating the crankcase

13 Dismantling the engine/gear unit: removing the crankshaft assembly

1 When the crankcases have been separated, the crankshaft will remain in the left-hand component because it is a shrink fit in the left-hand main bearing. Before heat is applied, the six bolts and washers that secure the left-hand oil seal retainer must be removed so that the retainer, complete with oil seal and 'O' ring can be detached.

2 Heat the left-hand crankcase around the main bearing housing to approximately 300 - 350°F (150 - 180°C), using the same technique and observing the same precautions as mentioned in the previous Section. When the correct temperature has been reached, a light tap or two on the end of the crankshaft should secure its release from the crankcase, complete with main bearing.

14 Dismantling the engine/gear unit: removing the gearbox components

1 Remove the fourth gear idler pinion from the end of the gearbox mainshaft.

2 Remove the thrust washer and kickstart pinion from the kickstart shaft.

3 Cut the locking wire and unscrew and remove the neutral position detent plunger from the crankcase.

Note: Before removing any parts from the selector and gear cluster assembly, make a sketch of the position of each gear and selector fork, as an aid to correct reassembly.

4 Remove the gear cluster assembly complete with the selector drum and selector forks.

5 Detach the selector drum and selector forks from the gear cluster and separate the mainshaft and layshaft assemblies. Remove the selector forks from the selector fork spindle.

6 Remove the gear pinions from the mainshaft in the following order, noting that the fourth gear idler has already been detached:

Circlip and first gear slider, then from the **opposite** end of the mainshaft remove the sleeve gear, third gear slider, circlip and second gear idler.

Retain the mainshaft pinions and circlips together as a group.

Remove the gear pinions from the layshaft in the following order:

Fourth gear pinion, first gear idler, circlip, second gear slider,

12.3 Main assembly will remain within left-hand crankcase half

14.2 Remove thrust washer and kickstart pinion

14.3 Withdraw detent plunger assembly

14.4 Remove gear cluster complete with selector drum

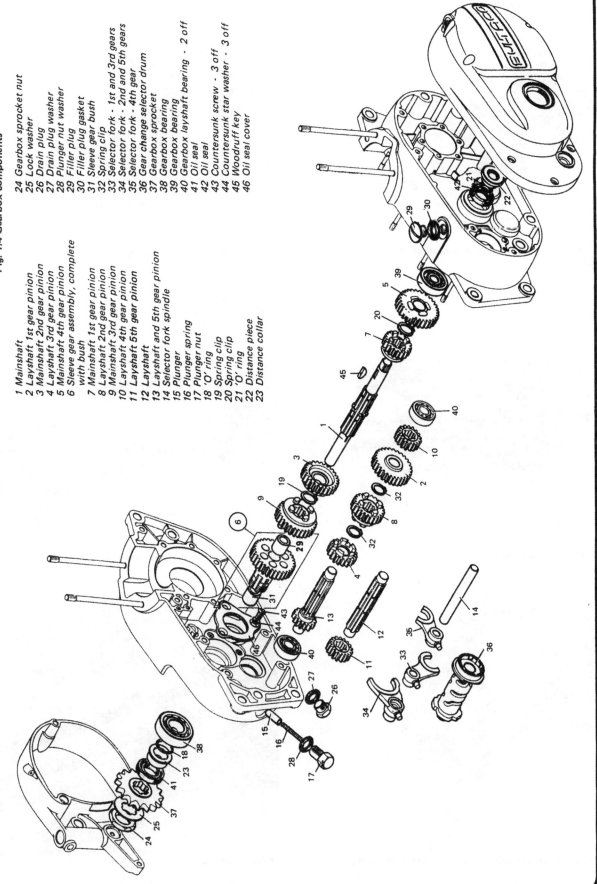

Fig. 1.4 Gearbox components

1 Mainshaft
2 Layshaft 1st gear pinion
3 Mainshaft 2nd gear pinion
4 Layshaft 3rd gear pinion
5 Mainshaft 4th gear pinion
6 Sleeve gear assembly, complete with bush
7 Mainshaft 1st gear pinion
8 Layshaft 2nd gear pinion
9 Mainshaft 3rd gear pinion
10 Layshaft 4th gear pinion
11 Layshaft 5th gear pinion
12 Layshaft
13 Layshaft and 5th gear pinion
14 Selector fork spindle
15 Plunger
16 Plunger spring
17 Plunger nut
18 'O' ring
19 Spring clip
20 Spring clip
21 'O' ring
22 Distance piece
23 Distance collar
24 Gearbox sprocket nut
25 Lock washer
26 Drain plug
27 Drain plug washer
28 Plunger nut washer
29 Filler plug
30 Filler plug gasket
31 Sleeve gear bush
32 Spring clip
33 Selector fork - 1st and 3rd gears
34 Selector fork - 2nd and 5th gears
35 Selector fork - 4th gear
36 Gear change selector drum
37 Gearbox sprocket
38 Gearbox bearing
39 Gearbox bearing
40 Gearbox layshaft bearing - 2 off
41 Oil seal
42 Oil seal
43 Countersunk screw - 3 off
44 Countersunk star washer - 3 off
45 Woodruff key
46 Oil seal cover

14.5 The gear cluster, prior to removal of the selector forks and rod

14.6 Remove circlip, followed by ...

14.6a ... first gear slider

14.6b Turn shaft over and remove sleeve gear, then ...

14.6c ...release circlip and draw off bearings, followed by ...

14.6d ...small cup

14.6e Pull off third gear slider ...

14.6f ... remove circlip and ...

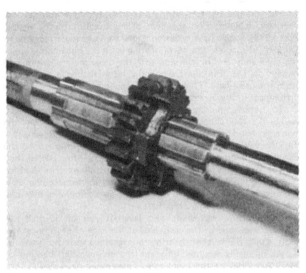
14.6g ... withdraw second gear idler

14.6h The layshaft assembly

14.6i Remove the first gear idler pinion. ...

14.6j ... then the circlip, followed by ...

circlip and third gear idler. The fifth gear pinion is integral with the shaft. Retain the layshaft gears and circlips together as a group.

7 If not previously removed, unscrew and remove the locknut and retaining nut securing one end of the kickstart return spring to the crankcase. Note the position of the shoulder of the retaining nut.

8 Temporarily fit the kickstart lever to its shaft, then turn the lever clockwise to remove the tension from the spring, so that the kickstart ratchet can be withdrawn. Turn the ratchet slightly to clear the hooked ratchet stop and pull the ratchet from the shaft, together with the ratchet spring and distance bush. Return the kickstart to the forward position and remove it from the shaft.

9 Remove the kickstart return spring from the kickstart shaft as described in Section 10.2 of this Chapter, then remove the shaft from the crankcase.

15 Dismantling the engine/gear unit: removing the crankshaft and gearbox main bearings

1 The crankshaft runs on either two or three journal ball bearings, depending on the year of manufacture of the engine. Post-1974 models have reverted to the two bearing arrangement. On 1973-4 models, the extra bearing is on the right-hand side of the crankshaft and in common with the other main bearings, will remain on the crankshaft when the latter has been withdrawn from the crankcase. This bearing arrangement necessitates some differences in the oil seal layout.

2 To remove the crankshaft main bearings, first use a soft, wedge-like tool shaped like a small cold chisel, to separate the two right-hand bearings (post 1973-74 three bearing models only). Tap the wedge gently between the bearings, working around the periphery to prevent the outer bearing from tilting. When they are approximately ¼ inch apart, a bearing puller can be used for the extraction of the outer bearing. Using the same puller, it is then comparatively easy to remove the two remaining bearings, one on each side of the crankshaft.

3 The gearbox mainshaft and layshaft run on journal ball bearings mounted in the crankcase at the ends of each respective shaft. Only the layshaft bearings are retained in blind housings.

4 An oil seal is mounted adjacent to each mainshaft bearing, on the outer sides of the crankcase. These must be removed first. On the right-hand side, the oil seal was removed whilst the clutch assembly was being dismantled. On the left-hand side, the oil

14.6k ...second gear slider, followed by ...

14.6L ...another circlip and the third gear idler

14.7 Locknut and retaining nut secure kickstart return spring

15.6 Heat crankcase to 400°F when removing main bearing

seal must be prised very carefully from the crankcase, and the flanged spacing bush and 'O' ring withdrawn.
5 Before the bearing itself can be removed, the bearing retainer must be taken off. This is held by three screws with washers.
6 Heat the crankcase halves to approximately 400°F (205°C) using a blow torch in the manner described previously, or a hot plate. When at the correct temperature, tap each crankcase half sharply on a block of wood, to shock the bearings out of position. Avoid damaging the jointing surfaces at all costs. If the bearings start to move, a hooked tool can be used to displace the layshaft bearings, which will prove the most difficult to remove. Take care to prevent damage to the bearing housings.

16 Examination and renovation: general

1 Before examining the parts of the dismantled engine for wear, it is essential that they should be cleaned thoroughly. Use a petrol/paraffin mix to remove all traces of old oil and sludge that may have accumulated within the engine and a cleansing agent such as Gunk or Jizer for the external surfaces. Special care should be taken when using these latter compounds, which require a water wash after they have had time to penetrate the film of grease and oil. Water must not be allowed to enter any of the interal oilways or parts of the electrical system.
2 Examine the crankcase castings for cracks or other signs of damage. If a crack is discovered, it will require specialist repair, or the renewal of both crankcases. Crankcases are supplied in matched pairs since it is considered bad engineering practice to renew only one. Under these latter circumstances there can be no guarantee that the main bearing housings have been bored exactly in line with one another.
3 Examine carefully each part to determine the extent of wear, if necessary checking with the tolerance figures listed in the Specifications section of this Chapter. The following sections of this Chapter describe how to examine the various engine components for wear and how to decide whether renewal is necessary.
4 Always use a clean, lint-free rag for cleaning and drying the various components prior to reassembly, otherwise there is risk of small particles obstructing the internal oilways.

17 Examination and renovation: main bearings and oil seals

1 Crankshaft main bearings in a two-stroke engine are particularly susceptible to damage caused by minute particles of dirt entering the engine through the carburettor air intake, especially if the air cleaner is ineffective or has been disconnected.
2 When the bearings have been removed from the crankshaft, wash them in a clean petrol/paraffin mix, dry them and then lubricate them with light oil. Since they are intended to be a shrink fit in the crankcases, they are manufactured with a slight clearance between the balls and the races and it is not possible to judge the amount of wear by checking clearances. The best test is to spin each bearing and check for abnormal noise or roughness as it slows down. Never spin a dry bearing. Play, if excessive, will be evident at the same time. If in doubt, renew the bearings. If they fail soon after reassembly, a further complete strip down will prove necessary.
3 Oil seals should be renewed as a matter of course. Apart from their oil retaining role, those fitted to the crankshaft also perform the role of an air seal, to prevent air leaking into the crankcase to dilute the incoming mixture as it enters the crankcase. Worn oil seals are one of the prime causes of difficult starting and uneven running. When fitting new oil seals, take extreme care, since their feather edges are very easily damaged.

18 Examination and renovation: crankshaft assembly

1 The crankshaft assembly operates under conditions of extreme stress and dimensional tolerances are critical. It is therefore essential to remedy crankshaft defects at a very early stage, to prevent more serious trouble developing.
2 Wash the complete flywheel assembly with a petrol/paraffin mix, to remove all surplus oil and combustion products. After drying, check the big end assembly for wear, then attempt to rock the top end of the connecting rod from side to side. If the amount of movement exceeds 0.098 in. (2.5 mm) the big end assembly is worn and must be reconditioned. Do not mistake side play for play at the top end of the rod because it is the vertical play that is all important, within reason. Side play should not exceed 0.008 to 0.010 in (0.25 mm).
3 Although it may be possible to run the engine for a brief period of service with a very small amount of play in the big end bearing, this action is not advised. Apart from the danger of the connecting rod breaking if the amount of wear should increase rapidly, a further complete engine strip will be required soon after reassembly to effect the reconditioning of the worn parts. It is best to effect a repair when the assembly is in any way suspect. Wear is denoted by a characteristic knock when the engine is running under load.
4 A factory reconditioned connecting rod and big end assembly is obtainable as a replacement, which will necessitate separation of the flywheels, replacement of the worn connecting rod and big end assembly and then realignment in a lathe to a very high standard of accuracy - a task beyond the means of most amateur mechanics. Reconditioning work of this nature should always be entrusted to a qualified expert and even then the chances are that the saving in cost, if any, will be only marginal.
Note: Big end assemblies are selectively fitted and it is advisable to purchase replacements only as a complete kit, comprising the connecting rod, small end and big end bearing assemblies and the crankpin. If a hollow crankpin is utilised, two crankpin expansion bushes are included in the kit.
5 If there is reason for doubt, crankshaft alignment must also be checked, a check best carried out by a qualified repairer. If the owner has access to and is able to use the necessary equipment, the following procedure is suggested: Mount the crankshaft in a lathe or on a pair of vee-blocks and a surface plate. Mount a dial guage just outside each flywheel with the gauge plunger in contact with the shaft. Turn the crankshaft a complete revolution and measure the run out towards the end of each shaft. If the run out exceeds 0.0004 in (0.010 mm) on the left-hand side or 0.0012 in (0.030 mm) on the right-hand side, the crankshaft must be reconditioned or exchanged. A poorly aligned crankshaft will absorb a surprising amount of power and give a very rough running engine.
6 If the engine has suffered a major crankshaft failure, the most probable cause is primary compression leakage from the crank chamber of the crankcase. Examine carefully both crankshaft oil seals and the crankcase jointing surfaces for evidence of the source of the leakage.

19 Examination and renovation: cylinder barrel

1 Bultaco cylinder barrels are fitted with a replaceable cast-iron liner. After long service or as the result of a piston seizure, the liner may be renewed, or bored and honed to accommodate an oversize piston. In either case, specialised equipment is needed, necessitating the assistance of a qualified repairer.
2 Give the cylinder bore a close visual inspection. If the surface is scored or grooved, indicative of an earlier seizure or a displaced circlip and gudgeon pin, a rebore is essential, regardless of the amount of bore wear. Compression loss will have a very marked effect on performance.
3 There will probably be a lip at the top end of the bore which marks the limit of travel of the top piston ring. The depth of the lip will give some indication of the amount of bore wear that has taken place, even though the amount of wear is rarely evenly distributed.
4 Check cylinder bore wear in relation to piston clearance. Invert the piston, less rings, so that it is upside down within the

19.10 If cylinder barrel has been rebored, round off port edges

cylinder barrel, positioned approximately one inch from the top of the barrel and with the long skirt facing the front of the cylinder. Check that the gudgeon pin bosses are in line with the axis of the crankshaft so that the piston is in its normal running position, although inverted. Using a narrow blade feeler gauge, measure the clearance between the front (long skirt) of the piston and the cylinder bore. If the clearance exceeds 0.006 in (0.16 mm) a rebore and the fitting of an oversize piston is necessary.

5 To check cylinder bore wear for taper, the bore diameter must be measured accurately at three different depths, using an inside micrometer positioned at right angles to the crankshaft axis. First position the micrometer just above the inlet port, next above the exhaust port and finally approximately ½ inch from the top of the cylinder. If the taper exceeds 0.0024 in (0.06 mm) rebore and hone the cylinder to the next oversize.

6 To check for cylinder ovality, the bore diameter must be measured with an internal micrometer first at right angles to the crankshaft axis, then in line with it, taking three readings in each case at the depths listed in the previous Section. If the out-of-round measurement exceeds 0.0016 in (0.04 mm) rebore and hone the cylinder to the next oversize. Pistons are available in four oversizes - 0.25, 0.50, 0.75 and 1.00 mm.

7 The clearance between a new cylinder and a new piston or a fully reconditioned assembly should be 0.0015 in (0.038 mm). At this clearance, it is necessary to run in the engine at moderate speed and load for at least two hours, during which time the rpm must not exceed 75% of that normally achieved. If a long running-in time is not possible, as for example just before a competition event, extra clearance must be provided - in this case 0.0025 in (0.063 mm) is permissible.

8 Under no circumstances modify or re-profile the ports in the search for extra performance. The size and location of the ports is critical in terms of engine performance and the dimensions chosen have been selected to give good, all-round performance consistent with a high standard of mechanical reliability.

9 Make sure the external cooling fins of the cylinder barrel are clean and not clogged with oil or dirt, which will otherwise impede the free flow of air and cause the engine to overheat. Remove any carbon that has accumulated in the exhaust ports with a blunt-ended scraper so that the surface of the ports is not scratched. Finish off with metal polish so that the ports have a smooth, shiny appearance. This will aid gas flow and prevent carbon from adhering so firmly on future occasions.

10 If the cylinder barrel has been rebored, it will be necessary to round off the extreme edges of the ports, to prevent rapid wear of the piston rings. Use a scraper or a hand grinder, taking off only the very minimum necessary and finishing off with fine emery cloth.

20 Examination and renovation: piston and piston rings

1 Attention to the piston and piston rings can be overlooked if a rebore is necessary because new replacements will be fitted.

2 If a rebore is not considered necessary, the piston should be examined closely. Reject the piston if it is badly scored, or discoloured as the result of the exhaust gases by-passing the rings.

3 Remove the piston rings by spreading the ends sufficiently with the thumbs to permit each to be lifted clear of the piston. This is a very delicate operation that must be handled with great care. Piston rings are brittle and will break very easily.

4 If the rings are stuck in their grooves, or have become gummed in by oily deposits, it is sometimes possible to free them by working small strips of tin along the back, to give a 'peeling' action. Again, great care is necessary to avoid breakage.

5 Remove all carbon from the piston crown, then finish off using metal polish. The resultant smooth, shiny surface will ensure that carbon will not adhere so readily in the future. Never use emery cloth for this purpose.

6 Examine the piston carefully for hairline cracks at the top edges of the transfer cutaways and in the vicinity of the gudgeon pin bosses. If tiny cracks are found, the piston must be renewed.

7 Check that the gudgeon pin bosses are not worn, or the circlip grooves damaged in any way. Check also that the piston ring pegs have not worked loose. They must locate the rings positively, to keep their ends from the being trapped in the ports.

8 The piston ring grooves may have become enlarged with use. The clearance between each ring and its groove must not exceed 0.006 inch side float.

9 Check each piston ring for wear by inserting it in the cylinder bore from the top and using the crown of the piston to push it down about 1½ inches so that it lies square in the bore. If the gap between the ends exceeds 0.013 inch of either ring, renew the pair as a set.

10 Examine the working surfaces of each ring. If discoloured areas are evident, the rings should be renewed since the patches denote blow-by of gas. Check also that there is no build-up of carbon on the back of the rings, a common two-stroke malady.

11 It cannot be over-emphasised that the condition of the piston and the piston rings in a two-stroke engine is of prime importance, especially since they control the opening and closing of the ports in the cylinder barrel by providing an effective seal. A two-stroke engine has only three basic working parts, one of which is the piston. It follows that the efficiency of the engine is very dependent on the condition of this component and the parts with which it is closely associated.

12 If the engine has been rebored and a new piston fitted, the amount of oversize should be clearly stamped on the piston crown. When fitting new piston rings, they must be of matching oversize. There is also a + (plus) or a — (minus) sign to ensure correct matching of the piston to the cylinder barrel whenever selective fits are employed.

21 Examination and renovation: small end bearing

1 Remove the caged needle roller bearing assembly from the small end of the connecting rod, then clean the assembly in a petrol/paraffin mix and dry if off. Using a magnifying glass, examine the needle bearing assembly for cracks at the corners of the slots in the bearing cage and for cracks or imperfections in the rollers themselves.

2 Clean the bearing surfaces of the small end eye of the connecting rod, and examine carefully for scratches, roughness or other signs of damage. If any damage is apparent, the connecting rod must be renewed, which in turn involves dismantling the complete flywheel assembly.

3 Examine the gudgeon pin for signs of wear. Lightly lubricate the pin and the needle bearing assembly and reassemble them in the small end. The gudgeon pin must be a good sliding fit without evidence of any play. Slide the pin backwards and forwards when checking for radial play. If play is evident, the gudgeon pin must be renewed, and possibly the connecting rod.

4 Examine each piston thrust washer for hairline cracks. Renew both washers if any are discovered, or if there are other signs of damage. Note that some of the earlier pistons have much thicker washers; if the piston is changed, it may be necessary to use different thickness washers, to match.

22 Examination and renovation: cylinder head

1 It is unlikely that the cylinder head will require any special attention apart from removing the carbon deposit from the combustion chamber. Finish off with metal polish; a polished surface will reduce the tendency for carbon to adhere and will also help improve the gas flow.

2 Ensure that the cooling fins are not obstructed and that they receive the full air flow. A wire brush provides the best means of cleaning.

3 Check the condition of the thread where the spark plug is inserted. The thread in an aluminium alloy cylinder head is damaged very easily if the spark plug is overtightened. If necessary, the thread can be reclaimed by fitting what is known as a Helicoil insert. Most agents have facilities for this type of repair, which is not expensive.

4 If the cylinder head joint has shown signs of oil seepage when the machine was in use, check whether the cylinder head is distorted by laying it on a sheet of plate glass. Severe distortion will necessitate a replacement head but if the distortion is only slight it is permissible to wrap some emery cloth (fine grade) around the sheet of glass and run down the joint using a rotary motion, until it is once again flat. The usual cause of distortion is uneven tightening of the cylinder head nuts and bolts. This is why the special technique detailed in Sections 5.10 and 7.1 of this Chapter should always be adopted.

5 There is no gasket at the cylinder head to cylinder barrel joint. If the jointing surfaces have shown a tendency to weep oil, it is permissible to grind the joint in after withdrawing the fixed studs temporarily, using find grinding paste. Grind with a semi-rotary motion, turning backward and forward. Raise the cylinder head occasionally and locate it in a different position before continuing the grinding operation. This will help distribute the grinding paste more evenly. Do not grind more than is necessary to achieve a good mating surface at both joints and make sure all traces of the grinding paste are removed, for this compound is hightly abrasive.

23 Examination and renovation: crankcase castings

1 Inspect the crankcase for cracks or any other signs of damage. If a crack is found, specialist treatment will be required to effect a satisfactory repair.

2 Clean off the jointing faces, using a rag soaked in methylated spirit to remove old gasket cement. Do not use a scraper because the jointing surfaces are damaged very easily. A leak-tight crankcase is an essential requirement of any two-stroke engine. Check also the bearing housings, to make sure they are not damaged. The entry to the housings should be free from burrs or lips.

3 Do not forget to check also the generator cover and the primary drive cover. Good jointing surfaces are essential especially in the case of the generator cover that has no intermediate gasket.

24 Examination and renovation: gearbox components

1 Examine carefully the gearbox components for signs of wear or damage such as chipped or broken teeth on the gear pinions and kickstart mechanism, worn or rounded dogs on the ends of the gear pinions, bent or worn selector forks, weakened or damaged springs and worn splines. If there is any doubt about the condition of a suspect part, it is preferable to play safe at this stage and renew the part concerned. Remember that if a suspect part should have to fail later, it will be necessary to strip the entire engine/gear unit yet again. Use new circlips and regrind ends to fit, if gear trains are dismantled.

2 It is advisable to renew the kickstart return spring as a matter of course, irrespective of whether it appears to be in good condition. This spring is in constant use and although it can be renewed without too much dismantling, at a later stage, it is preferable to do so now.

3 As mentioned earlier, the gearbox oil seals should be renewed whilst the engine unit is dismantled, and the gearbox bearings, if there is evidence of any play or roughness.

25 Examination and renovation: clutch actuating mechanism

1 It is unlikely that this mechanism will require much attention. Check that the operating arm moves quite freely within its housing, is well greased and that it has not become a particularly slack fit. The 'O' ring and felt seal at the top must be in good condition.

2 If it is necessary to dismantle the operating arm assembly in order to grease the shaft, first mark the position of the operating arm on the splined end of the shaft, so that it can be replaced in an identical position. Remove the nut and bolt that secures the arm, and lift it off, followed by the felt washer and 'O' ring seal. The shaft is retained within the outer case by a shouldered bolt fitted on the outside of the housing. When this bolt is removed, the shaft can be withdrawn for greasing.

3 On late models the operating arm is fitted to the top of the cover. Earlier models have the arm on the underside of the cover.

4 Do not forget to oil the folding kickstart arm otherwise it will seize.

26 Examination and renovation: clutch assembly

1 Because the clutch is of the metal to metal variety, there is not such a high rate of wear on the individual clutch plates as on the more conventional type with friction linings. If clutch slip has caused the clutch plates to be suspect, the most likely cause is because they have become polished. They can be reclaimed by repunching the existing indentations. It also helps if the plates are reassembled in different order, so that the same areas do not rub against each other.

2 Check the tongued edges of the clutch plates both inner and outer for burrs or indentations and the slots in the clutch centre and the clutch housing. Often, indentations will be made by the tongues of the clutch plates after a very lengthy period of service and the tongues will become trapped in these, making the clutch plates difficult to separate and initiating clutch drag. It is permissible to file the burrs off the edges of the tongues and to dress the slots of the clutch assembly with a file until the sides of the slots are square once again. But this can only be carried out if the amount of damage is small, otherwise clutch chatter will set in and greatly increase the rate of wear.

3 The clutch push rod must be straight and slide quite freely within the hollow gearbox mainshaft. Check also the hardened ends, to ensure they are not worn or breaking up. The need for frequent clutch adjustment is often a sign of a push rod with softened ends that wear quite rapidly when any loading is applied.

4 Line up all the clutch springs and check that they are all of identical length. If some have compressed more than others, renew the whole set. It is a wise precaution to compare the length with a new, unused spring. Springs that have compressed with use will sooner or later give rise to clutch slip.

27 Examination and renovation: primary chain

1 The primary chain is of the endless type and runs in ideal conditions, having the lower run immersed in oil and the tension kept constant by means of a spring-loaded tensioner. In consequence, it is likely to require attention for a considerable period of use, apart from an examination to ensure there are no cracked or broken rollers or sideplates.

2 Chain tension can be checked only when the chain tensioner is in position. As there is no means of adjustment for wear, the chain will have to be renewed if excessive play develops. With full tension applied by the tensioner, there should not be more than 3/8 inch play measured in the middle of the upper run.

3 If the chain has to be renewed, check the condition of the engine and clutch sprocket too. If they are badly worn or hooked, they too should be renewed on the same occasion. It is always preferable to renew the whole transmission system as a set.

28 Examination and renovation: gear selector mechanism

1 The gear selector mechanism comprises what is known as a trigger shaft - a shaft with two spring loaded pawls at the outward facing extremity. These pawls engage with indentations in the end of the gear selector drum, to form the positive stop for individual gear selection. The trigger lever, which fits over the end of the shaft, has a peg, which engages with a slot in an arm attached to the gearchange lever shaft. The latter is spring-loaded with a hairpin-shaped return spring, so that it will always return to the same position.

2 Check that the pawls move freely in their housings and that the springs have not weakended or broken. Check also that the hairpin return spring is in good condition; renew it if there is any doubt.

29 Reassembly: general

1 Before the engine, clutch and gearbox components are reassemblied, they must be cleaned thoroughly so that all traces of old oil, sludge, dirt and gaskets are removed. Wipe each part clean with a dry, lint-free rag to make sure that there is nothing to block the internal oilways of the engine.

2 Lay out all the spanners and other tools likely to be required so that they are close at hand during the reassembly sequence. Make sure the new gaskets and oil seals are available - there is nothing more infuriating then having to stop in the middle of a reassembly sequence because a gasket to some other vital component has been overlooked.

3 Make sure the reassembly area is clean and unobstructed and that an oil can with clean engine oil is available so that the parts can be lubricated before they are reassembled. Refer back to the torque wrench settings and clearance data where necessary. Never guess or take a chance when this data is available.

4 Do not rush the reassembly operation or follow the instructions out of sequence. Above all, do not use excess force when parts will not fit together correctly. There is invariably good reason why they will not fit, often because the wrong method of assembly has been used.

30 Reassembling the engine/gear unit: replacing the kickstart shaft and spring*

1 Reverse the procedure detailed in Section 10, paragraphs 7 to 9 of this Chapter. Tension the return spring by turning the kickstart shaft a half turn, then sliding the ratchet onto the shaft.

2 Fit the locknut to the shoulder nut and tighten it fully.

31 Reassembling the engine/gear unit: replacing the crankshaft assembly

1 Rest the left-hand crankcase inner face uppermost across two pieces of wood and heat it to approximately 350°F (180°C) with a blowlamp or other localised source of heating. When it is at the correct temperature, lower the crankshaft assembly into position so that the left-hand main bearing enters its housing squarely and can be pressed home. No undue force should be necessary; a few light taps on the right-hand end of the crankshaft should suffice, using a raw hide mallet.

2 Allow the crankcase to cool down, then check that the assembly revolves quite freely, without any trace of roughness. The main bearing must be fully home in its housing. Check that the assembly is central in the crankcase casting.

32 Reassembling the engine/gear unit: replacing the gear cluster

1 If the gear trains have been dismantled, reassemble them again and try a mock assembly with both shafts, and the selectors, on the workbench, to ensure replacement in the crankcase will be in the correct order.

2 Install the complete gear cluster in the left-hand crankcase, after it has been allowed to cool down, with the exception of the mainshaft fourth gear. Refit the kickstart gear pinion on the kickstart shaft, followed by the thrust washer, then refit the mainshaft fourth gear.

3 Slide the selector drum into its housing. Install the three selector forks, commencing with the third/fifth gear fork and checking to ensure the cam following pin engages with the correct groove in the drum. Follow up with the first/second gear fork, the shoulder of which will fit into the counterbore of the fork already fitted. Then fit the fourth gear fork with its shoulder facing upwards. It is possible to install it in the opposite direction, but under these circumstances, the gear change will not function.

4 Slide the selector fork rod into position and check that it pushes home into its recess.

5 Fit the detent plunger assembly, again checking to ensure the selector drum is in the neutral position. Refit the locking wire so that the assembly cannot work loose.

6 Before proceeding further, check that all the shafts revolve quite freely. The crankcase can now be joined together.

32.1 Attempt mock assembly on bench first

32.2 When inserting the sleeve gear pinion, grease to prevent oil seal damage

32.3 Slide the selector drum into its housing and engage with selectors

32.4 Press selctor rod into housing

32.5 Ensure detent plunger engages with neutral notch

34.1 Install pawls as shown

34.1a Rotate, to ensure pawls are in depressed position

Chapter 1: Engine, clutch and gearbox

34.2 Use new gasket when refitting

35.1 Don't omit 'O' ring seal

35.2 Grease flanged spacer prior to insertion

33 Reassembling the engine/gear unit: joining the crankcase halves

1 Before the right-hand crankcase can be fitted, it must be heated to 350°F, so that the right-hand main bearing will enter its housing without risk of damage.
2 Ensure both crankcase alignment dowels are replaced in the left-hand crankcase and that a new gasket is fitted. This gasket must be fitted dry. Check that the gasket will contact the cylinder base gasket when the latter is fitted, otherwise there is risk of loss of crankcase compression. Preferably the gasket should extend above the joint, so that it can be trimmed off afterwards.
3 Lower the right-hand crankcase into position, making sure the right-hand main bearings enter their housing squarely. Do not use force and if the crankcase does not align correctly the first time, remove it and start all over again. When the two crankcase halves meet satisfactorily, give the right-hand crankcase a few light taps with a raw hide mallet to make sure it has seated correctly, then fit star washers to each stud projecting from the right-hand crankcase, followed by the nuts, which should then be tightened. Note that there is also a Nyloc nut, and Allen screw and star washer that have to be fitted to the top left-hand side of the front portion of the crankcase. Allow the crankcase assembly to cool completely, replace the bearing retainers and their screws, then check that all the shafts revolve quite freely, including the crankshaft.

34 Reassembling the engine/gear unit: replacing the gear selector pawls

1 Install the pawls as shown in the accompanying photograph, then press them both inwards whilst rotating the trigger lever so that they are retained in the permanently depressed condition by the shoulder on the trigger shaft.
2 Fit the pawl cover, using a new gasket, pushing the trigger shaft into the centre of the gear selector drum. Then rotate the trigger lever clockwise until it has moved through 90° and points to the top of the crankcase.
3 Replace the three screws that secure the pawl cover in position, using Loctite to keep them in place.
4 Refit the gear selector shaft, which passes through the crankcase. Make sure the slot in the operating arm has engaged correctly with the peg of the trigger lever. Fit the return spring.

35 Reassembling the engine/gear unit: replacing the clutch and primary drive

1 Replace the 'O' ring seal on the right-hand end of the crankshaft, then insert the oil seal complete with retainer plate and second 'O' ring after first greasing the inside bore of the oil seal to ensure it is not damaged. Fit and tighten the retainer plate screws, six in total.
2 Grease and insert the flanged spacer, taking special care to avoid damaging the oil seal during insertion. Refit the Woodruff key in the right-hand end of the crankshaft.
3 Grease and insert the flanged spacer over the end of the gearbox mainshaft, taking care not to damage the oil seal.
4 Assemble as a unit the clutch centre and outer housing, the primary chain and the engine sprocket cum external flywheel assembly. They must be assembled as a complete unit because the chain is of the endless type. Check that the keyway in the engine sprocket assembly is in line with the key in the end of the crankshaft.
5 When the complete assembly has positioned itself correctly, fit a new star washer under the clutch centre nut, refit the nut and tighten to a torque setting of 75 lb/ft, whilst holding the clutch steady with the clutch holding tool.
6 Without removing the clutch holding tool, refit the engine

35.3 Don't forget spacer on gearbox mainshaft

35.4 Reassemble clutch, flywheel and chain in unison

35.7 Stick ball bearing to clutch end piece with heavy grease

35.7a Use clutch spring compressing tool to replace pins

35.8 Refit chain tensioner and spring

35.9 No need to tighten clutch adjustment nut at this stage

sprocket retaining nut and tighten this to an identical setting of 75 lb/ft. Remove the holding tool and wire the sprocket nut as previously. There is a felt washer behind the sprocket over the sleeve gear, which must be in first class condition and well greased.

7 Reassemble the clutch by alternating the plates, finishing with a plate that engages with the clutch centre. Take the clutch end piece (mushroom) and stick the ball bearing to the inner end with heavy grease. Insert the assembly into the hollow gearbox mainshaft, the refit the clutch pressure plate, spring caps and springs. Use the clutch spring compressing tool to compress each spring in turn, so that the retaining pins can be fitted. Fit the shallow cup below each pin first, so that the recess faces upwards. When correctly installed, each pin should fit within the recess of each cup.

8 Refit the chain tensioner and its spring, so that it presses on the underside of the lower run of the chain.

9 Recheck the security of all components in the primary drive, making sure the adjuster screw and locknut are in the centre of the clutch pressure plate but not at this stage tightened. Check that the outer cover locating dowels are in position, then refit the right-hand outer cover, using a new gasket at the joint (no gasket cement). Refit and tighten the eight Allen screws that retain the over in position. They are of differing lengths and must be replaced in their original locations - noted during the earlier dismantling sequence. Note there is an oil seal in the cover through which the end of the gearchange shaft passes. Care is needed because the shaft has serrated ends; grease the shaft to obviate the risk of damage to the oil seal. A felt washer should then be inserted - between the cover and the back of the gearchange lever, when fitted.

36 Reassembly of the engine/gear unit: refitting the flywheel magneto

1 Turn the engine unit over so that the left-hand crankcase now faces upwards and the whole unit stands firmly on the workbench. Fit the left-hand oil seal, 'O' ring and bearing retainer assembly, replace and tighten the six retaining screws.

2 Replace the stator plate assembly, noting that the rubber grommet through which the electrical leads pass slots into the forward facing part of the crankcase. Align the stator plate with the marks made previously, then replace and tighten the three retaining screw around the periphery.

3 Place the Woodruff key in the left-hand end of the crankshaft, locate with the flywheel rotor and lower the latter into position on the crankshaft taper, pushing it well home. Hold the rotor with the clutch holding tool and tighten the retaining nut to 75 lb/ft. Do not omit the spring washer located below this nut.

4 Before refitting the flywheel magneto rotor, check that it is free from metallic particles, which will adhere to it as a result of its magnetism. Also check the condition of the contact breaker points because the rotor will have to be removed again should it be found they require attention at a later stage.

5 Check the ignition timing as described in Chapter 3.

37 Reassembling the engine/gear unit: replacing the final drive sprocket

1 Grease and slide the long clutch push rod through the hollow gearbox mainshaft after checking that the ball bearing on the end of the clutch end piece (mushroom) has not been misplaced.

2 Replace the sprocket over the splined end of the sleeve gear, noting that there is an 'O' ring, shouldered spacer and oil seal that precede it (and an oil seal retainer secured by three screws on some models). These components may have been fitted earlier, as detailed in Section 33.3 of this Chapter.

3 Refit the tab washer and then the nut that retains the final drive sprocket. Hold the sprocket firmly with a piece of chain or with a chain spanner, then tighten the nut to a setting of 75 lb/ft. Bend over the tab washer and fit the small circle of felt around the push rod to act as a seal.

4 Before fitting the outer cover (use a new gasket at joint) check that the clutch actuating mechanism in the cover is quite free. Check also that the outer cover locating dowels are in position and that the fibre spacer washer is stuck to the rearmost 'pillar' extending from the crankcase, with thick grease. Replace and tighten the retaining screws.

38 Reassembling the engine/gear unit: replacing the piston, cylinder barrel and cylinder head

1 Stand the engine unit upright and pad the mouth of the crankcase with clean rag. Reinsert the small end needle roller bearing, then refit the piston, warming it if necessary, to ease the insertion of the gudgeon pin. Fit new circlips, never the originals, which should always be discarded. Make sure they engage correctly with their respective grooves. The piston crown is stamped with an arrow, which must face forwards in all cases. If the piston is acidentally reversed, there will be a remarkable loss of power, if indeed the engine will run at all.

2 Make sure the correct spacing washers are used between the piston bosses and each end of the small end bearing. The piston should have minimum end float. If a new piston is fitted, it may be necessary to fit matching washers.

3 The piston rings are of the Dykes type, being 'L' shaped in cross section. They must be replaced in their correct positions with the wide ring at the top - astride the piston ring end pegs and with the thinner portion extending upwards from each piston ring groove, as in the accompanying photograph.

4 Fit a new cylinder base gasket, after trimming off any excess of the crankcase joint gasket that may protrude through the joint gap. No gasket cement is necessary.

5 It is preferable to use a piston ring clamp even though the edge of the cylinder barrel has a tapered lead-in. Oil the bore, then compress both rings and gently lower the cylinder barrel down the holding down studs, making sure the exhaust stub points to the front of the machine. This task is best accomplished with a second pair of hands, one person compressing the rings and the other manipulating the cylinder barrel. When the rings enter the bore, slide the barrel downwards a short distance, then remove the rag from the crankcase mouth before it is seated at the joint with the crankcase.

6 When the cylinder barrel is seated, refit the cylinder head, replacing the retaining nuts and bolts in their original positions, with the exception of those to which the head steady is attached. Tighten them finger tight only at this stage. The engine unit is now ready for replacement in the frame.

39 Replacing the engine/gear unit in the frame

1 Lift the engine/gear unit into the frame from the right-hand side, tilting it initially to clear the front frame lug. When it is in position, refit and tighten the nuts and bolts that retain the engine to the built-in lugs on the frame.

2 This task is made much easier if a second person is available, one to lift in the engine unit whilst the other steadies the frame.

3 Replace the cylinder head steady, then tighten the cylinder head nuts and bolts in the sequence given earlier. Tighten nuts 1 to 6 in numerical sequence to a torque setting of 2 lb/ft and continue tightening in this order with an increase of 2 lb/ft at each stage until nuts 5 and 6 reached a final setting of 11 lb/ft and nuts 1 to 4, 14 lb/ft. Make sure the nut and bolt securing the other end of the head steady to the frame is also tight.

4 Reconnect the electrical leads at the terminal box on the front down tube. Their colour coding should obviate the risk of wrong connections. Replace the spark plug lead and supressor cap at the same time.

5 Replace the carburettor on the induction stub or flange. Tighten the flange bolts, or if the carburettor is of the stub fitting variety, the hose clips. In the case of the former, do not

35.9a Felt washer goes behind gearchange lever

36.2 Grommet fits into recess within crankcase casting

36.3 Locate flywheel rotor with key in crankshaft end

36.3a Tighten to recommended torque setting

37.4 Stick fibre washer with grease before cover is fitted

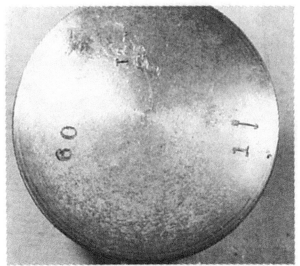
38.1 Arrow must face forwards. Note selective fit markings on crown

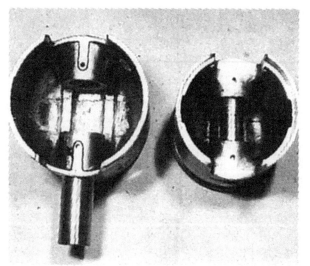
38.2 Distance between piston bosses is likely to vary

38.3 Piston rings of Dykes type are fitted

38.3a Rings must locate astride ring pegs of piston

38.4 New cylinder base gasket is essential

39.1 Lift engine unit in from right-hand side

39.1a Sherpa front engine mounting

Chapter 1: Engine, clutch and gearbox 37

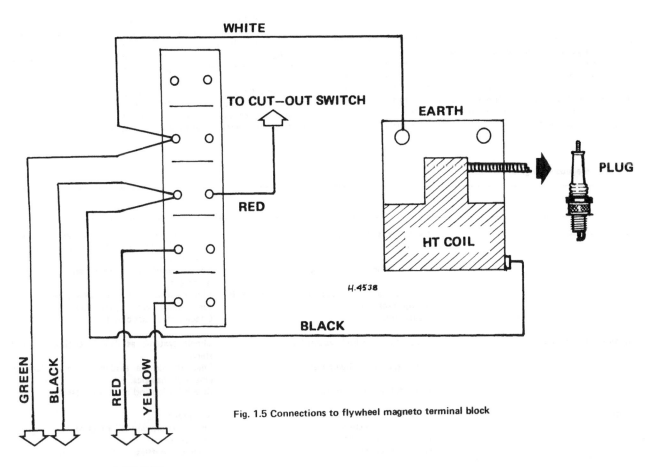

Fig. 1.5 Connections to flywheel magneto terminal block

39.1b Sherpa rear engine mountings

overtighten, or there is risk of bowing the flange joint, to create an air leak.

6 Refit the complete exhaust system, by reversing the dismantling procedure. Make sure all the springs are tensioned and located correctly, to prevent the system working loose.

7 Reconnect the air cleaner and, where necessary, replace the side panels. Reconnect the clutch cable and adjust the clutch so that there is approximately 3/16 inch free play at the handlebar lever. Coarse adjustment is by means of the nut in the centre of the clutch pressure plate, which has been left loose for this purpose. When adjustment is approximately correct, tighten the locknut whilst holding the screw to prevent it from turning. Replace the slotted screw in the outer cover, which provided access. Make final adjustments with the adjuster on the end of the handlebar lever.

8 Refit the kickstart and gearchange levers, making sure the oil seal precedes the former in the housing provided. Set them at their correct angles on the splined shafts, previously marked, before tightening the pinch bolts. Note the kickstart leans forward, at a somewhat unusual angle.

9 Check that all the drain plugs have been replaced and tightened, then add the recommended quantities of lubricant - 300 ccs of SAE 30* oil for the primary drive case and 600 ccs of SAE 90 oil for the gearbox.

10 Replace the petrol tank and reconnect the carburettor, also
*SAE 5 or 10, Sherpa T models only.

the underside of the two halves of the tank. Do not omit the bolt that retains the front of the tank to the frame. Replace the bolts in the seat fairing, or those used to retain the nose of the seat.

40 Starting and running the rebuilt engine

1 When the initial start-up is made, run the engine slowly for the first few minutes, especially if the engine has been rebored. Check that all the controls function correctly and that there are no oil leaks before taking the machine on the road.
2 Remember that a good seal between the piston and the cylinder barrel is essential for the correct functioning of any two-stroke engine. In consequence a rebored engine will require more careful running-in than its four-stroke counterpart. There is a far greater risk of engine seizure during the first hundred miles if the engine is permitted to work hard.
3 Do not add extra oil to the petrol/oil mix in the mistaken belief that it will aid running-in. More oil means less petrol and the engine will run with a permanently weakened mixture, causing overheating and a far greater risk of engine seizure. Keep to the recommended proportions.
4 Do not tamper with the exhaust system or run the engine without the baffles fitted to the silencer. Unwarranted changes in the exhaust system will have a very noticeable effect on engine performance, invariably for the worst.

41 Fault diagnosis - engine

Symptom	Cause	Remedy
Engine will not start	Defective spark plug	Remove plug and lay on cylinder head. Check whether spark occurs when engine is kicked over.
	Dirty or closed contact breaker points	Check condition of points and whether gap is correct.
	Air leak at crankcase or worn oil seals around crankshaft	Flood carburettor and check whether mixture is reaching the spark plug.
	Clutch slip	Check and adjust clutch.
Engine runs unevenly	Ignition and/or fuel system fault	Check systems as though engine will not start.
	Blowing cylinder head gasket	Leak should be evident from oil leakage where gas escapes.
	Incorrect igniton timing	Check timing and reset if necessary.
Lack of power	Incorrect ignition timing	See above.
	Fault in fuel system	Check system and filler cap vent.
	Blowing head gasket	See above.
	Choked silencer	Clean out baffles.
High fuel/oil consumption	Cylinder barrel in need of rebore and o/s piston	Fit new rings and piston after rebore.
	Oil leaks or air leaks from damaged gaskets or oil seals	Trace source of leak and replace damaged gaskets or seals.
Excessive mechanical noise	Worn cylinder barrel (piston slap)	Rebore and fit o/s piston.
	Worn small end bearing (rattle)	Replace bearing and gudgeon pin.
	Worn big end bearing (knock)	Fit new big end bearing.
	Worn main bearing (rumble)	Fit new journal bearings and seals.
Engine overheats and fades	Pre-ignition and/or weak mixture	Check carburettor settings. Check also whether plug grade correct.
	Lubrication failure	Is correct measure of oil mixed with petrol?

42 Fault diagnosis - clutch

Symptom	Cause	Remedy
Engine speed increases but machine does not respond	Clutch slip	Check clutch adjustment for pressure on pushrod. Also free play at handlebar lever. Check condition of clutch plates, also free length of clutch springs. Replace if necessary.
Difficulty in engaging gears. Gear changes jerky and machine creeps forward, even when clutch if fully withdraw	Clutch drag	Check clutch adjustment for too much free play.
	Clutch plates worn and/or clutch drum	Check for burrs on clutch plate tongues or

Chapter 1: Engine, clutch and gearbox

	Clutch assembly loose on mainshaft	indentations in clutch drum slots. Dress with file. Check tightness of retaining nut. If loose, fit new tab washer and retighten.
Operating action stiff	Damaged, trapped or frayed control cable	Check cable and replace if necessary. Make sure cable is lubricated and has no sharp bends.
	Bent pushrod	Replace.

43 Fault diagnosis - gearbox

Symptom	Cause	Remedy
Difficulty in engaging gears	Gear selectors not indexed correctly	Reassemble with detent plunger in neutral notch of selector drum. Check also position of positive stop pawls.
	Gear selector forks bent	Replace.
	Broken or misplaced selector springs	Replace broken springs and re-locate as necessary.
Machine jumps out of gear	Worn dogs on ends of gear pinions	Replace worn pinions.
	Detent plunger or pawls stuck	Free plunger assembly or pawls
Kickstarter does not return when engine is turned over or started	Broken or badly tensioned kickstarter return spring	Replace spring or retension.
Gear change lever does not return to normal position	Broken return spring	Replace.

Chapter 2 Fuel system and lubrication

Contents

General description ... 1	Carburettor: dismantling and examination ... 7
Petrol/oil mix: correct ratio ... 2	Carburettor: checking the settings ... 8
Petrol tank: removal and replacement ... 3	Air cleaner, removal, cleaning and replacement ... 9
Petrol tap: removal and replacement ... 4	Exhaust system: cleaning ... 10
Petrol feed pipes: examination ... 5	Fault diagnosis: fuel system and lubrication ... 11
Carburettor: removal ... 6	

Specifications

Fuel tank capacity

Alpina models	2.1 Imp. gallons, 2.5 US gallons, 9.5 litres
Frontera models	2.4 Imp. gallons, 2.9 US gallons, 11 litres
Pursang models	1.5 Imp. gallons, 1.85 US gallons, 7.0 litres
Sherpa models	1 Imp. gallon, 1.2 US gallons, 4.5 litres

Carburettor

Bultaco model	Alpina		Frontera		Pursang			Sherpa	
Capacity ccs	250	350	250	360	200	250	360	250	350
Make of carburettor	.... Amal								
Type No.	627	627	2036	2036	932	L-1036-11	L-1036-11	627	627
Choke size mm	27	27	36	36	38	38	38	27	27
Slide	3	3	2.5	2.5	2.5	2.5	2.5	3.5	3.5
Needle notch	2nd	2nd	4th	3rd	2nd	—	—	2nd	2nd
Needle jet	106	106	108	108	107	107	107	106	106
Main jet	160	160	360	370	320	370	390	150	150
Pilot jet	30	20	25	30	30	35	35	20	20
Choke jet (starter)	—	—	50	50	—	—	—	—	—

Lubricant

Two-stroke oil	Mixing ratio 25 : 1
SAE 40 motor oil	Mixing ratio 20 : 1

1 General description

The fuel system comprises a fuel tank from which a petrol/oil mixture of controlled proportions is fed by gravity to the float chamber of an Amal Concentric carburettor. A petrol tap, with an integral gauze filter, is located beneath the rear end of the fuel tank, which has provision for turning on a small reserve quantity of fuel when the main supply is exhausted. There are additional nylon mesh filters in the main feed to the float chamber of the carburettor, and surrounding the main jet.

The carburettor is not fitted with a choke for cold starting. The float chamber has a tickler which, when depressed, will lower the float assembly and temporarily enrich the mixture. This technique is necessary only when the engine is cold and the air temperature is low. This does not, however, apply to the Frontera and latest Pursang models, which have an Amal MkII Concentric carburettor fitted with a choke plunger. This plunger is operated in the vicinity of the carburettor and not from the handlebars.

2 Petrol/oil mix: correct ratio

1 Because the engine relies on the petroil system for its lubrication, a measured quantity of oil must always be added to the petrol. If a self-mixing two-stroke oil is used, the correct ratio is 180 ccs of oil to each Imperial gallon of petrol (25 : 1). Mix the oil and petrol together by shaking them vigorously, if the mix is not delivered by a self-mixing pump. Do not use petrol with an octane rating lower than 90.

2 Although the use of a two-stroke, self-mixing oil is advised, standard motor oil of SAE 40 rating can be used as a substitute. In this instance, the mix ratio is 230 ccs of oil to a gallon of petrol (20 : 1) with even more emphasis on the mixing routine.

Chapter 2: Fuel system and lubrication

The use of motor oil has two major disadvantages. Firstly, the oil is more inclined to settle out if the machine is left standing for any period of time and it will be necessary to wiggle the handlebars vigorously before the first start-up, in order to redisperse the oil. Secondly, the oil contains certain additives which will ash when burnt and give rise to plug fouling. Plus whiskering is one of the side effects.

3 Only a mineral oil should be used. Vegetable-base oils are quite unsuitable since they mix with petrol only sparingly and will cause the internal oilways to block with sludge if they mix with traces of the remaining mineral oil.

4 It will be realised that the lubrication of the engine is dependent solely on the intake of the petrol/oil mixture from the carburettor. In consequence, it is not advisable to coast the machine down a long hill with the throttle closed, otherwise there is risk of seizure through the temporary lack of lubricant.

5 The gearbox has its own separate supply of lubricating oil and is quite separate from the engine in this respect. Two stroke mixing oils must NEVER be used in in the gearbox.

3 Petrol tank: removal and replacement

1 Detach the petrol feed pipe from the petrol tap by pulling it off the push on connection. Then either drain the tank by opening the tap(s) and allowing the fuel to drain into a suitable receptacle, or keep the tap(s) closed and pull off one end of the tube that passes under the nose of the tank to join the two halves. This acts as the alternative drain-off point.

2 Remove the two (or four) screws that hold the seat fairing (Alpina and Sherpa models) and the nut, washer and rubber washer that retains the nose of the tank to the frame tube. The tank will now lift off, together with the seat.

3 Take care to store the tank in a safe place, away from naked flames. Petrol vapour is both highly explosive and highly inflammable, which underlines the need for special precautions.

4 Petrol tap: removal and replacement

1 It is rarely necessary to remove the petrol tap (two taps, Frontera and Pursang models) unless the integral gauze filter has become blocked to the extent that the flow of petrol is impeded. Before the tap(s) can be removed the petrol tank must first be drained of fuel completely.

2 The tap(s) will unscrew from the base of the fuel tank by placing a spanner across the hexagon provided for this purpose. Note that the tap seats on a housing containing a rubber washer, the latter of which should always be renewed when it is disturbed. All jointing surfaces must be perfectly clean, to obviate leaks.

3 The gauze is easily cleaned once the tap is removed. If, however, the tap itself leaks around the turn-on lever, or if the lever gets either very slack or very tight, the tap itself will have to be renewed.

4 Some models have a different type of tap in which the tap lever can be dismantalled for renewal of the sealing washers. Remove the two screws that retain the lever cover, and the whole assembly can be dismantalled. Repeat, if necessary, for the second tap. The tank must be drained off first.

5 Replace by screwing the tap(s) back into the underside of the tank and then replace the fuel tank by reversing the dismantalling procedure. Tighten the nut at the front of the tank sufficiently to begin compressing the rubber washer. Refit the various fuel lines, refill the fuel tank and check for leaks.

5 Petrol feed pipes: examination

1 Plastic fuel pipes, reinforced with nylon cord, are used throughout. Of robust construction, they are unlikely to require attention, other than an occasional inspection to ensure there are

2.2 Screws hold rear of tank/seat fairing on Sherpa models

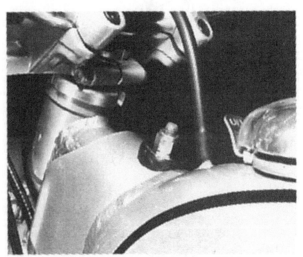
2.2a Nut, bolt and rubber washer retain nose of tank

4.4 Remove both screws to detach tap lever assembly

Chapter 2: Fuel system and lubrication

no cracks or splits, especially in the vicinity of the points of connection.

2 At each point of connection, the push-on joint should be secured by means of a stout metal clip and screw. Never use wire, even as a temporary substitute as this will cut into the plastic pipe and eventually cause it to fail at this point.

3 If a pipe has to be trimmed to remove damage at one end, make sure it is not cut too short so that the radius of any bend is so tight that it restricts the flow.

6 Carburettor: removal

1 All models employ the new MkII Amal Concentric carburettor, the very latest models having a special angular version that incorporates a choke plunger for easy starting in cold weather. Carburettors of either the stub or flange mounting type are fitted, the method of attachment depending on the model of machine. All carburettors are basically similar in design and operating principle.

2 To detach the carburettor, disconnect the fuel line(s) from the banjo union at the float chamber, or alternatively remove the banjo union nut, then the union itself. In the latter case, take special care of the nylon gauze filter, which forms part of the banjo union assembly and is easily damaged.

3 Loosen the hose clip that retains the carburettor to the cylinder barrel inlet (stub fitting type) or unscrew and remove the two bolts and washers on the flange mounting (flange fitting type). Pull off the air cleaner hose, after slackening the hose clip.

4 Before removing the carburettor from the machine, unscrew the two screws in the top of the carburettor and lift off the top, together with the throttle cable, slide and needle assembly. The latter can be taped out of harms way to some convenient part of the machine until examination is required at a later stage. The carburettor fitted to most of the later models has a screw-on top.

7 Carburettor: dismantling and examination

1 To remove the float chamber, unscrew the two crosshead screws on the underside of the mixing chamber. The float chamber can then be pulled away complete with float assembly and sealing gasket. Remove the gasket and lift out the horseshoe-shaped float, float needle and spindle on which the float pivots.

2 When the float chamber has been removed, access is available to the main jet, jet holder and needle jet. The main jet threads into the jet holder and should be removed first, from the underside of the mixing chamber. Next unscrew the jet holder which contains the needle jet. The needle jet cannot be removed until the jet holder has been unscrewed and removed from the mixing chamber because it threads into the jet holder from the top. There is no necessity to remove the throttle stop or air adjusting screws. Note that the carburettor has an extra gauze filter cap over the main jet, which will pull off.

3 Check the float needle for wear which will be evident in the form of a ridge worn close to the point. Renew the needle if there is any doubt about its condition, otherwise persistent carburettor flooding may occur.

4 The float itself is unlikely to give trouble unless it is punctured and admits petrol. This type of failure will be self-evident and will necessitate renewal of the float.

5 The pivot needle must be straight - check by rolling the needle on a sheet of plate glass.

6 It is important that the gasket between the float chamber and the mixing chamber is in good condition if a petrol tight joint is to be made. If it proves necessary to make a replacement gasket, it must follow the exact shape of the original. A portion of the gasket helps retain the float pivot in its correct location; if the pin rides free it may become displaced and allow the float to rise, causing continual flooding and difficulty in tracing the cause. Use Amal replacements whenever possible.

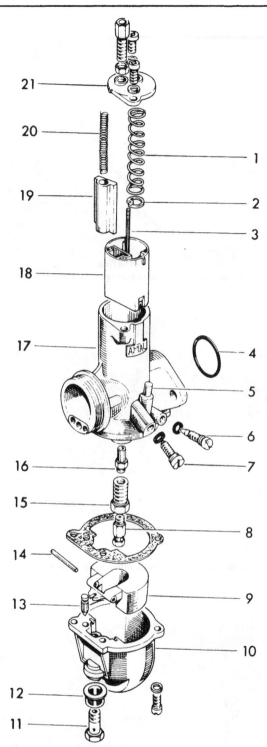

Fig. 2.1 Component parts of the Amal Concentric carburettor

1 Throttle return spring
2 Needle clip
3 Needle
4 'O' ring
5 Tickler
6 Pilot jet screw
7 Throttle stop screw
8 Main jet
9 Float
10 Float chamber
11 Banjo union bolt
12 Filter
13 Float needle
14 Float hinge
15 Jet holder
16 Needle jet
17 Mixing chamber body
18 Throttle valve (slide)
19 Air slide (choke)
20 Air slide return spring
21 Mixing chamber top

7.1 Remove the two screws on the underside of the mixing chamber

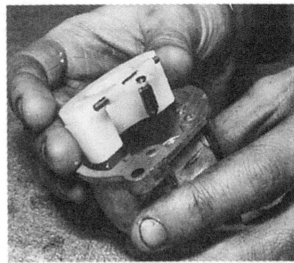

7.1a Float needle is minute and easily lost

7.2 Access is now available to the jet assembly

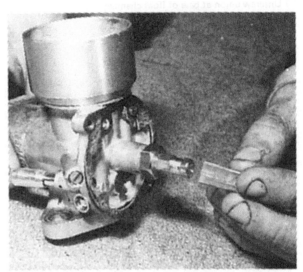

7.3 Detachable nylon gauze surrounds main jet

7.3a Jet assembly will unscrew from mixing chamber

7.5 Float pivot needle and gasket must be in good condition

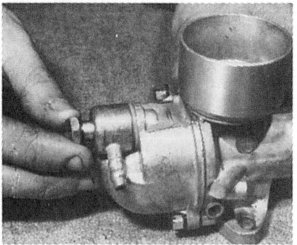

7.7 Unscrew union at base of float chamber ...

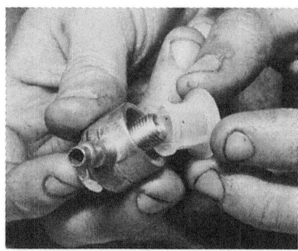

7.7a ... to expose nylon filter

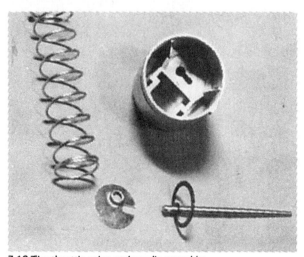

7.12 The throttle valve and needle assembly

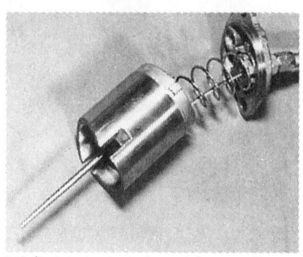

7.12a Check the throttle valve for wear before reassembly

7 Remove the union at the base of the float chamber and check that the inner nylon filter is clean. All sealing washers must be in good condition.

8 Make sure that the float chamber is clean before replacing the float and float needle assembly. The float needle must engage correctly with the lip formed on the float pivot; it has a groove that must engage with the lip. Check that the sealing gasket is placed OVER the float pivot spindle and the spindle is positioned corectly in its seating.

9 Check that the main jet and needle jet are clean and unobstructed before replacing them in the mixing chamber body. Never use wire or any pointed instrument to clear a blocked jet, otherwise there is risk of enlarging the orifice and changing the carburation. Compressed air provides the best means, using a tyre pump if necessary.

10 Before refitting the float chamber, check that the jet holder and main jet are tight. Do not invert the float chamber, otherwise the inner components will be displaced as the retaining screws are fitted. Each screw should have a spring washer to obviate the risk of slackening.

11 When replacing the flange mounting carburettor, check that the O-ring seal in the flange mounting is in good condition. It provides an airtight seal between the carburettor flange and the cylinder flange to ensure the mixture strength is constant. Do not overtighten the carburettor retaining nuts for it is only too easy to bow the flange and give rise to air leaks. A bowed flange can be corrected by removing the O-ring and rubbing down on a sheet of fine emery cloth wrapped around a sheet of plate glass, using a circular motion. A straight edge will show if the flange is level again, when the O-ring can be replaced and the carburettor refitted.

12 Before the carburettor top is replaced, check the throttle valve for wear. A worn valve is often responsible for a clicking noise when the throttle is opened and closed. Check that the needle is not bent and that it is held firmly by the clip.

13 The choke plunger assembly fitted to the latest models should not require attention, provided the lever and plunger move quite freely. Although the carburettor has a more angular appearance, it is very similar in construction to the standard Concentric design.

14 Reassemble the carburettor in the reverse order of dismantling. Note that flange fitting carburettors must have a heat insulator between the carburettor flange and the induction manifold, to reduce the transfer of heat from the cylinder barrel.

8 Carburettor: checking the settings

1 The various sizes of the jets and the throttle slide, needle and needle jet are predetermined by the manufacturer and should not require modification. Check with the Specifications list if there is any doubt about the values fitted.

2 Slow running is controlled by a combination of the throttle stop and pilot jet settings, irrespective of the type of carburettor fitted. Commence by screwing inwards the throttle stop screw so that the engine runs at a fast tick-over speed. Adjust the pilot jet screw until the tick-over is even, without either misfiring or hunting. Screw the throttle stop outwards again until the desired tick-over speed is obtained. Check again by turning the pilot jet screw until the tick-over is even. Always make these adjustments with the engine at normal working temperature and remember that the characteristics of a two-stroke engine are such that it is very difficult to secure an even tick over at low engine speeds. Some prefer the engine to stop when the throttle is closed completely, but in a two-stroke engine with petroil lubrication there is always risk of oil starvation if the machine is coasting with the throttle closed.

3 As a rough guide, up to 1/8th throttle is controlled by the pilot jet, from 1/8th to ¼ throttle by the throttle slide cutaway, from ¼ to ¾ throttle by the needle position and from ¾ to full throttle by the size of the main jet. These are only approximate divisions; there is a certain amount of overlap.

4 The normal setting for the pilot jet screw is approximately one and a half full turns out from the fully closed position. If the engine 'dies' at low throttle openings, suspect a blocked pilot jet.

5 Guard against the possibility of incorrect carburettor adjustments that result in a weak mixture. Two-stroke engines are very susceptible to this type of fault, which will cause rapid overheating and subsequent engine seizure. Some owners believe that the addition of a little extra oil to the petrol will help prolong the lift of the engine, whereas in practice quite the opposite occurs. Because there is more oil the petrol content is less and the engine runs with a permanently weakened mixture!

9 Air cleaner: removal, cleaning and replacement

Alpina models

1 To gain access to the air cleaner assembly, remove the right-hand side cover located just below the nose of the seat. It is retained by a single screw and spring.

2 Remove the filter element lid and withdraw the filter element and the metal gauze screen. Wash both in petrol to remove all dust and dirt, then dry with compressed air.

3 Soak the filter element in clean SAE 30 oil and let it drain. Coat the inside surface of the air cleaner box with grease. Insert the metal gauze, then the filter element, after wringing it out and wiping the surface of excess oil. Refit the filter element lid, then replace the side cover and tighten the securing screw.

Frontera models

4 On this model it is first necessary to remove both side covers from under the nose of the seat. Each is retained by three screws.

5 Unscrew the tensioner that holds the air cleaner element in place by means of three springs. Remove the element and clean it with a blast of compressed air. If badly contaminated, wash in petrol first.

6 Soak the element in SAE 30 oil, wring out the excess and wipe the surface of the foam. Replace the unit in the reverse order of dismantling.

Pursang models

7 Use a similar technique to that detailed for the Frontera models, as under the previous heading. In this instance the filter

9.9 Wire clips holds filter element and metal screens in place

9.11 Don't forget to check condition of hose to carburettor intake

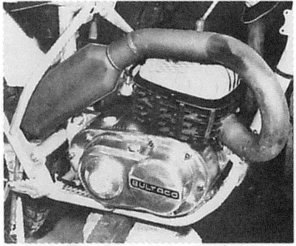
10.1 Sherpa exhaust system is of the upswept variety

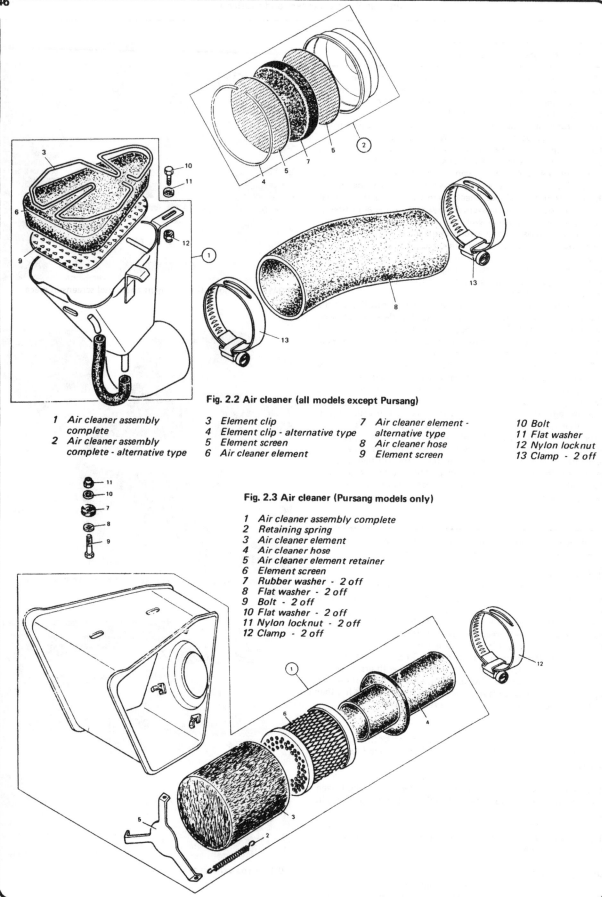

Fig. 2.2 Air cleaner (all models except Pursang)

1 Air cleaner assembly complete
2 Air cleaner assembly complete - alternative type
3 Element clip
4 Element clip - alternative type
5 Element screen
6 Air cleaner element
7 Air cleaner element - alternative type
8 Air cleaner hose
9 Element screen
10 Bolt
11 Flat washer
12 Nylon locknut
13 Clamp - 2 off

Fig. 2.3 Air cleaner (Pursang models only)

1 Air cleaner assembly complete
2 Retaining spring
3 Air cleaner element
4 Air cleaner hose
5 Air cleaner element retainer
6 Element screen
7 Rubber washer - 2 off
8 Flat washer - 2 off
9 Bolt - 2 off
10 Flat washer - 2 off
11 Nylon locknut - 2 off
12 Clamp - 2 off

Chapter 2: Fuel system and lubrication

10.1a ... silencers are of unusual shape

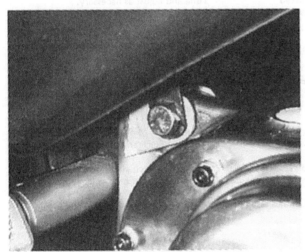

10.3 Upper engine unit bolt retains first silencer

10.3a Rearmost silencer is attached to frame stay

10.2 Sherpa system is in two parts, held together with springs

element should be cleaned with compressed air only, and **not** oiled.

Sherpa models

8 Before the air cleaner element can be removed, it is first necessary to remove the seat and petrol tank, as a complete unit.
9 A wire spring holds the metal screen in position. Press the ends of the spring together so that it can be lifted out of position, together with the metal screens - one at the top and one at the bottom.
10 Remove the nylon filter element and wash both the element and the metal screens in petrol, then blow them dry with compressed air. Wipe the air cleaner unit then coat the inside with grease. Replace the lower metal screen and the nylon element, then coat the top of the latter with a spray-on lubricant such as that used for chains, which will not clog the pores. Replace the upper metal screen and the wire clip, and then refit the combined seat and petrol tank unit.
11 Check to ensure the connecting rubber to the carburettor intake is not split or cracked. This is important, as apart from the risk of ingress of dust into the engine, a leak will seriously upset the carburation. The carburettor is specially jetted to take into account the presence of the air cleaner. For similar reasons, never run the machine with the air cleaner disconnected, or without the element in place. A badly weakened mixture will result, which may cause irreparable engine damage.
12 The air cleaner unit must be serviced at the prescribed intervals, otherwise performance will suffer.

10 Exhaust system: cleaning

1 Exhaust systems can be divided into two groups, the upswept, tucked-in type fitted to the Alpina and Sherpa models, and the downswept crossover type, fitted to the Frontera and Pursang models. The former utilises two quite separate silencers of unusual shape; the latter is tapering and has a large bore silencer fitted after the expansion box.
2 The upswept system is made in two parts, which are held together by stout springs and the downswept system is one-piece. Both systems are retained to the exhaust port by three springs. To release the springs, pull them off with a hook made from a piece of stout wire.
3 The rear end of the system is retained by a bolt in the vicinity of the gearbox, which may also act as one of the engine/gear unit retainers. The rearmost silencer of the upswept system is also retained by a bolt which passes through a lug on the frame loop that retains the rear mudguard. The downswept system is retained near the extremity by a long stay attached by bolts to a clip around the silencer or tail pipe. The Frontera models have a

protective underplate attached to the lowest part of the system.
4 Because the exhaust gases of a two-stroke are of an oily nature, the exhaust system will require cleaning out at regular intervals, to obviate the risk of a build-up of oily sludge causing back pressure. If the system has been left for too long, so that the build-up is heavy, it may be necessary to burn out the carbon with a blowlamp, and then repaint the system with heat resisting black paint to prevent it from rusting. Many cases of poor performance can be traced directly to a blocked or partially blocked exhaust system that is causing a back pressure.
5 Under no circumstances tamper with the system in any way, or replace it with another of different design. The system is designed to blend in with the characteristics of the engine and give good performance with the desired reduction in noise level. More noise is in no way indicative of greater speed, and quite often the reverse. If the machine is used in competition, an increase in noise level may cause the rider to be 'booked' at a meeting where noise meter tests are made, which could ultimately lead to his suspension.
6 When replacing the exhaust system, check that it is securely fitted and that all the springs are correctly located and the bolts tight. An exhaust system that works loose during a competition represents a hazard not only to the rider himself, but others who may be following or passing.

11 Fault diagnosis: fuel system and lubrication

Symptom	Cause	Remedy
Excessive fuel consumption	Air cleaner choked or restricted	Clean or if paper element fitted, replace.
	Fuel leaking from carburettor. Float sticking	Check all unions and gaskets. Float needle seat needs cleaning.
	Badly worn or distorted carburettor	Replace.
	Carburettor incorrectly adjusted	Tune and adjust as necessary.
	Incorrect silencer fitted to exhaust system	Do not deviate from manufacturer's original silencer design.
Idling speed too high	Throttle stop screw in too far. Carburettor top loose	Adjust screw. Tighten top.
Engine does not respond to throttle	Back pressure in silencer. Float displaced or punctured	Check baffles in silencer. Check whether float is correctly located or has petrol inside.
	Use of incorrect silencer or baffles missing	See above. Do not run without baffles.
Engine dies after running for a short while	Blocked air hole in filler cap	Clean.
	Dirt or water in carburettor	Remove and clean out.
General lack of performance	Weak mixture; float needle stuck in seat	Remove float chamber or float and clean.
	Air leak at carburettor joint or in crankcase	Check joints to eliminate leakage.
Excessive white smoke from exhaust	Too much oil in petrol, or oil has separated out	Mix in recommended ratio only. Mix thoroughly if mixing pump not available.

Chapter 3 Ignition and lighting systems

Contents

General description ... 1	Rectifier: general description ... 11
Flywheel magneto: checking the output ... 2	Ballast resistor: function ... 12
Ignition coil: checking ... 3	Headlamp: replacing bulbs and adjusting beam height ... 13
Contact breaker: adjustment ... 4	Stop and tail lamp bulb: removal and replacement ... 14
Contact breaker points: removal, renovation and replacement ... 5	Stop lamp switch: adjustment ... 15
	Speedometer lamp: replacement ... 16
Ignition timing: checking and resetting ... 6	Horn: alternative types ... 17
Condenser: removal and replacement ... 7	Wiring: layout and inspection ... 18
Spark plug: checking and resetting the gap ... 8	Lighting and handlebar switches: general ... 19
Lighting systems: general ... 9	Fault diagnosis: ignition system ... 20
Battery: examination and maintenance ... 10	Fault diagnosis: lighting systems ... 21

Specifications

Contact breaker gap ... 0.013 - 0.017 (0.35 - 0.45 mm) all models except Pursang

Ignition timing btdc
Alpina models	2.5 - 2.7 mm
Frontera models	2.7 - 2.9 mm
Pursang 200cc	2.7 - 2.9 mm
250cc	2.6 - 2.8 mm
360cc	2.2 - 2.4 mm
Sherpa models	2.8 - 3.0 mm

Spark plugs
Size	14 mm
Reach	¾ inch
Gap	0.013 - 0.017 inch (0.35 - 0.45 mm)

Spark plug recommendations

Bultaco model	Alpina		Frontera	
Capacity	250	350	250	360
Lodge	CLNY	3HLN	RL49	3HLN
KLG	FE30	FE100	FE260	FE100
Champion	N12Y	N3	N57R	N3

Bultaco model	Pursang	Sherpa	
Capacity	200/250	360	250/350
Lodge	RL49	RL49(F) 3HLN (R)	CLNY
KLG	FE280	FE280(F)	FE30
Champion	N58R	N58R (F) N3 (R)	N12Y

(F) front plug
(R) rear plug

Electrical system
Flywheel magneto

Bultaco model	Alpina		Frontera
Capacity	250	350	250/350
Make	FEMSA	FEMSA	FEMSA

Chapter 3: Ignition and electrical systems

	Alpina		Frontera
Type No.	VAR41-51	VAR41-52	VAR41-51
Voltage (lighting)	............6 volts............		
Ignition coil			
Type No.	............BA9-53............		
Battery (6 volt)			
Type No.	............BT16-1............		
Rectifier			
Type No.	............CRB2-1............		
Bulb ratings			
Headlamp	........25/25W........		..30/30W.
Tail/stop	........5/15W........		..5/15W.
Pilot lamp (parking)	nil		....5W.

Flywheel magneto

Bultaco model	Pursang			Sherpa	
Capacity	200	250	360	250	350
Make	Motoplat	Motoplat	FEMSA	FEMSA	FEMSA
Type No.	96003030-1	96003030-1	GEA-11	VAR41-51	VAR41-52
Voltage (lighting)	N/A	N/A	N/A	........6 volts........	
Ignition coil					
Type No.	—	—	—	........BA9-53.........	
Bulb ratings					
Headlamp	N/A	N/A	N/A	........25/25W........	
Tail/stop	N/A	N/A	N/A	........5/15W.........	
Pilot lamp (parking)	N/A	N/A	N/A	............nil............	

Note: All bulbs 6 volt rating.

1 General description

1 The electrical energy needed to power the ignition system and in some cases an additional lighting system is derived from a flywheel magneto mounted on the left-hand end of the crankshaft. The flywheel magneto generates alternating current as it revolves, which is fed to a separate ignition coil and converted into high tension current necessary to produce the spark across the spark plug points. The precise moment at which this spark occurs is determined by a contact breaker assembly behind the flywheel rotor, or in the case of the Pursang models, by a special electronic system that obviates the need for a mechanical 'make and break'.

Where there is an additional lighting requirement, a lighting coil is included in the flywheel magneto assembly to provide the current for the extra electrical load. The system may, or may not, include a battery; if it does, a rectifier is an additional requirement to convert the alternating current into direct current for charging the battery. Irrespective of whether or not a battery is included, a ballast resistor is used to control the flow of current, which would otherwise tend to surge under certain conditions and overload parts of the lighting system.

The system requires very little maintenance, since the flywheel magneto has no rubbing parts that are subject to wear. The ignition timing has to be set and maintained to a high standard of accuracy for optimum performance, a task simplified by the use of an electronic system in the Pursang range, obviating the need to clean and adjust contact breaker points on a routine maintenance basis.

2 Flywheel magneto: checking the output

1 There is no satisfactory method of checking the electrical output from the flywheel magneto without the use of sensitive multi-meter test equipment. Since the average owner is unlikely to have access to this equipment or instruction in its use, the testing of equipment considered to be faulty is a task for a Bultaco dealer or an auto electrical expert.

3 Ignition coil: checking

1 The ignition coil is a sealed unit attached to a metal plate, which bolts to a lug on the top frame tube, immediately behind the steering head gusset. To gain access, it will be necessary to remove the fuel tank.

2 In the event of failure or malfunctioning, the ignition coil is not repairable and will have to be renewed as a complete unit. Before suspecting the ignition coil, however, check that the condenser in the contact breaker circuit is not at fault. Refer to Section 6 of this Chapter for the relevant details. Condenser breakdown will cause the ignition system to fail or malfunction, giving the impression that the ignition coil may be at fault.

3 To check the coil, remove the spark plug from the cylinder head and lay it across the cylinder head fins with the lead still attached, so that the outside body of the plug is earthed via the fins. If the ignition system is functioning in a satisfactory manner, an intense blue spark should jump across the spark plug points as the engine is turned over on the kickstarter. If there is no spark at all, check again with a replacement plug, in case the original is faulty.

4 Contact breaker: adjustment

All models, except Pursang

1 The contact breaker assembly is attached to the stator plate of the flywheel magneto. Limited access for checking purposes can be gained through a small 'window' cut in the flywheel rotor, or in some cases two windows. If the points require attention, however, it is essential to withdraw the flywheel rotor first, as described in Chapter 1, Section 8.

2 To check the gap at the contact breaker points, rotate the flywheel rotor until a window is opposite the points and by rocking the rotor backwards and forwards, check that they are in the fully open position. If there are two windows, it may be necessary to use an adjacent window.

3 Insert a feeler gauge and measure the gap between the points. If the gap is correct, it should be within the range 0.013 - 0.017 inch (0.35 - 0.45 mm). If outside this range, adjustment is required.

4 It is possible to adjust the points without removing the flywheel rotor, even though the task requires some care on account of the limited access. Above the points are two screws, one larger than the other. Slacken the larger screw, then turn the smaller screw in the direction desired, to either open out or close

Chapter 3: Ignition and electrical systems

2.1 Electrical connections are made at terminal block

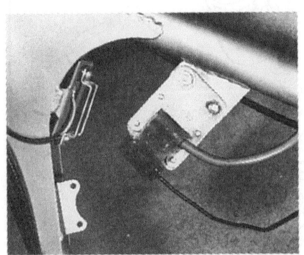

3.1 Ignition coil is attached to plate on frame tube, under tank

5.1 Flywheel rotor must be pulled off for full access to points

up the gap. The small screw is eccentric, to facilitate adjustment. When the gap is correct, tighten the larger screw again, then re-check the gap. The points should now be set correctly.

5 Hold the points open and examine the faces of the contacts. If they are blackened, pitted or burnt, they require further attention, in which case the flywheel rotor must be withdrawn. See next Section.

6 Remember that a contact breaker gap that is too small will retard the ignition setting, or if too large, will advance it. When checking the accuracy of the ignition timing, as described later in this Chapter, the contact breaker gap must always be correctly set first.

Pursang models only

7 The Pursang models have an electronic ignition system which obviates the need for contact breaker points. In consequence, this and the following Section can be ignored entirely as far as those models only are concerned. NEVER detach the plug lead when the engine is running, especially at high speeds. This will cause irrepairable damage to the ignition system.

5 Contact breaker points: removal, renovation and replacement

1 To gain full access to the contact breaker points, first withdraw the flywheel rotor, as described in Chapter 1, Section 8.

2 If the points have a blackened or burnt appearance, or have pitted badly, they will have to be removed for dressing with an oil stone. If it is necessary to remove a substantial amount of material, however, before the surfaces can be restored, it is preferable to renew the complete points assembly and discard the originals.

3 To detach the points assembly, remove the screw that holds the two electrical leads, then the screw that retains the assembly to the flywheel magneto stator plate. The moving contact point can be released from its pivot by prising off the circlip that retains it, and removing the return spring.

4 Dress the points with an oil stone or fine emery cloth. Keep them absolutely square during the dressing operation, or they will make angular contact on reassembly and will quickly burn away.

5 Replace the contact by reversing the dismantling procedure, not forgetting to lightly grease the pivot shaft of the moving contact and to place a few drops of light oil in the felt pad that bears on the internal cam of the flywheel rotor. It is attached to the moving contact and helps lubricate the cam so that the fibre pad at the tip of the contact arm does not wear too rapidly. Check that the contact faces are clean and free of oil.

6 Replace the flywheel rotor, then re-set the contact breaker points to the recommended gap, as described in the preceding Section.

6 Ignition timing: checking and resetting

All models except Pursang

1 If the ignition timing is correct, the contact breaker points will commence to separate when the piston is the following amount before top dead centre:

Alpina models	2.5 - 2.7 mm
Frontera models	2.7 - 2.9 mm
Sherpa models	2.8 - 3.0 mm

A high degree of accuracy is essential if optimum performance is to be achieved, hence the timing should always be set very carefully.

2 The piston position can be verified by using Bultaco Service Tool 132974, a vernier gauge that will screw into the hole normally occupied by the spark plug. A plunger will rest on the piston crown to read off the piston position on a scale, alongside a moveable scale that can be pre-set with a thumb screw.

3 To check the exact moment at which the points separate,

52

Chapter 3: Ignition and electrical systems

Fig. 3.1 Ignition system
1 T iron models only
2 Magneto assembly complete
3 Femastronic or Motoplat unit (choice depends on model)
4 High tension coil (ignition coil)
5 Femastronic or Motoplat spark unit
6 Magneto cover
7 Magneto cover gasket
8 Magneto screw - 3 off
9 Spark plug cover
10 Rotating magnet
11 Rotating magnet
12 Magneto washer
14 Contact breaker base plate
15 Lighting coil
16 Lighting coil
17 Low tension coil
18 Low tension coil
19 Screw - 4 off
20 Coil screw - 4 off
21 Coil washer - 4 off
22 Condenser
23 Condenser
24 Contact breaker screw - 2 off
25 Washer
26 Terminal screw
27 Eccentric adjusting screw
28 Eccentric adjustment screw
29 Gasket spacing washer
30 Contact breaker points
31 Contact breaker points
32 Star washer - 2 off
33 Grommet
34 Washer
35 Wire with terminal
36 Washer
37 Magneto retaining nut
38 Shouldered magneto retaining nut - alternative
39 Lubricating felt
40 Felt bracket
41 Nut
42 Washer
43 Bracket
44 Wire with terminal connections
45 Wire with terminals
46 Insulated connector
47 Wire protector
48 Tab
49 Earth wire
50 Timing pin
51 Bolt - 2 off
52 Nut - 2 off
53 Spring washer - 3 off
54 Spring washer
55 Spring washer
56 Allen screw - 3 off
57 Allen screw
58 Star washer - 4 or 6 off (depending on model)
59 Nylon locknut - 2 off
60 Cable strap - 2 off
61 Flat washer - 3 off
62 Woodruff key

connect a battery and test lamp wired in series so that one lead is earthed to the frame and the other is attached to the connection between the black and red wires in the plastic insulator clamped below the flywheel magneto. The bulb will remain alight until the points separate. A Bultaco timing light, which operates from the mains is also available for this purpose.

4 If the setting is not correct, it will be necessary to withdraw the flywheel rotor as described in Chapter 1, Section 8, and slacken the three screws around the periphery of the stator plate so that it can be moved in the direction required. Turning the plate clockwise will retard the setting and anti-clockwise advance it. Turn only a small amount at a time, then replace the flywheel and re-check. Often, as many as four attempts will prove necessary before the setting is correct.

5 As mentioned earlier, the contact breaker points must be correctly gapped before the ignition timing is checked or reset. Alterations made afterwards will affect the accuracy of the setting.

Pursang models only

6 The Pursang models have a Motoplat electronic ignition system and in consequence a different technique is necessary. It is recommended, however, that the Bultaco vernier gauge (Service Tool 132974) is used to determine the piston position if a high standard of accuracy is to be maintained.

7 There is a small hole in the flywheel rotor and if the ignition timing is correct, a pin inserted into this hole should align exactly with a hole in the upper left-hand corner of the stator plate behind the rotor. The timing is correct if the pin enters the hole when the piston is in the following position before top dead centre:

2.7 - 2.9 mm	200cc model
2.6 - 2.8 mm	250cc model
2.2 - 2.4 mm	360cc model

The shank of a twist drill of the appropriate size will act conveniently as a pin for the timing routine.

8 If the setting is not correct, it will be necessary to withdraw the flywheel rotor as described in Chapter 1, Section 8 and slacken the three screws around the periphery of the stator plate behind the flywheel. Then move the stator plate either backwards or forwards to advance or retard the setting (forwards - retard, backwards - advance). Replace the flywheel and recheck the setting. Note that it may be necessary to repeat this procedure up to four times until the desired standard of accuracy is obtained. Loose assembly of the flywheel should, however, aid the approximate alignment of the holes in the early stages.

9 On models fitted with a twin spark plug cylinder head, the vernier gauge should be inserted in the rearmost position.

7.3 Condenser is inserted into stator baseplate

7 Condenser: removal and replacement

All models except Pursang

1 A condenser is included in the contact breaker circuitry to prevent arcing across the contact breaker points when they separate and to help intensify the spark at the spark plug. It is connected in parallel with the points and if a fault develops, the ignition system will not function correctly.

2 If the engine proves difficult to start or if misfiring occurs, it is possible that the condenser has failed. To check, examine the surface of the contact breaker points. If they have a blackened and burnt appearance, this is an indication that arcing is taking place. If the engine can be persuaded to start, observe the points through the window in the rotating flywheel rotor. If more than just an occasional, intermittent spark occurs, the condenser has failed and must be renewed.

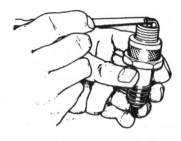
Checking plug gap with feeler gauges

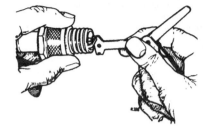
Altering the plug gap. Note use of correct tool

Fig. 3.2a Spark plug maintenance

White deposits and damaged porcelain insulation indicating overheating

Broken porcelain insulation due to bent central electrode

Electrodes burnt away due to wrong heat value or chronic pre-ignition (pinking)

Excessive black deposits caused by over-rich mixture or wrong heat value

Mild white deposits and electrode burnt indicating too weak a fuel mixture

Plug in sound condition with light greyish brown deposits

Fig. 3.2b Spark plug electrode conditions

Chapter 3: Ignition and electrical systems

3 It is not possible to test a condenser without the appropriate test equipment, but as a replacement is a low cost item, a check by substitution provides an alternative means of checking. To gain access to the condenser, withdraw the flywheel rotor by following the procedure given in Chapter 1, Section 8, the condenser and then the screw that retains the condenser body to the stator plate. Withdraw the condenser and insert the replacement. Make sure the connections are re-made tightly, especially the screw retaining the condenser body, since this must make a good earth connection. Replace the flywheel rotor, then the outer cover assembly.

8 Spark plug: checking and resetting the gap

1 A 14 mm long reach (¾ inch) spark plug is fitted to all the models in the Bultaco range covered by this manual. The 360cc Pursang models have a two-plug cylinder head, using an additonal 'soft' plug for warming up before the lead is transferred to the racing plug.
2 The gap at the plug points should be maintained within the range 0.013 - 0.017 inch (0.35 - 0.45 mm). Reset if necessary by bending the outer electrode, never the centre electrode, otherwise the insulator will crack and cause particles to drop into the engine whilst it is running.
3 The condition of the spark plug electrodes and insulator can be used as a reliable guide to engine operating conditions. See accompanying diagrams.
4 Always carry a spare spark plug of the correct grade. The plug in a two-stroke engine leads a hard life and is liable to fail more readily than when fitted to its four-stroke counterpart.
5 Never over-tighten a spark plug, otherwise there is risk of stripping the threads from the cylinder head, especially those cast in light alloy. A stripped thread can be repaired by using what is known as a 'Helicoil' insert, a low cost service that is operated by a number of dealers.
6 Use a spanner that is a good fit, otherwise the spanner may slip and break the plug insulator. The plug should be tightened sufficiently to seat firmly on its sealing washer.
7 Make sure the plug insulator cap is a good fit and free from cracks. This cap contains the suppressor that eliminates radio and TV interference.
8 See the Specifications heading at the beginning of this Chapter for spark plug grade recommendations. Always use the recommended grade of spark plug, or the exact equivalent in another manufacturer's range. Remember that two-stroke engines are particularly sensitive to unwarranted changes in sparking plug specification.

9 Lighting system: general

1 A lighting system is available as an optional extra for the Sherpa trials models and as part of the standard equipment for the Alpina and Frontera trial/enduro models. Two alternative systems are available, one of which utilises a battery, an essential requirement in some American States where more stringent lighting regulations have to be met.
2 The Sherpa trials models imported into the UK during recent years have had the extra lighting coil included in the flywheel magneto. Sammy Miller Equipment can supply the remainder of the lighting kit, including the wiring loom, with simple instructions for connecting up.
3 When a battery is included in the lighting system, an additional requirement is a rectifier, to convert the ac current from the flywheel magneto generator to dc for charging the battery. If the battery fails to hold its charge and is known to be in good condition, the rectifier is the component most likely to give trouble, assuming there is output from the lighting coil in the flywheel magneto.
4 Some wiring layouts include what is known as a ballast resistor, to prevent the electrical system from overload in the event of bulb failure, at high rpm.

10 Battery: examination and maintenance

1 When a battery is fitted, usually in the case of the Alpina models, it is contained in a compartment under the nose of the seat, usually on the left-hand side of the machine. To gain access, detach the respective side cover and remove the strap that retains the battery in position.
2 Maintenance is normally limited to keeping the electrolyte level just above the plates and separators. Modern batteries have translucent plastics cases, which make the check of electrolyte level much easier.
3 Unless acid is spilt, which may occur if the machine falls over, the electrolyte should always be topped up with distilled water until the correct level is restored. If acid is spilt on any part of the machine, it should be neutralised with an alkali such as washing soda and washed away with plenty of water, otherwise serious corrosion will occur. Top up with sulphuric acid of the correct specific gravity (1.260 - 1.280) ONLY when spillage has occurred.
4 It is seldom practicable to repair a cracked battery case because the acid already in the joint will prevent the formation of an effective seal. It is always best to replace a cracked battery, especially in view of the corrosion that will be caused by the leakage of acid.
5 The battery is probably the most neglected component on any motorcycle. Never leave it in a permanently run down condition, even if the lights are not used and the generator fails to charge it up. It should be removed from the machine and given a charge with a battery charger, at the rate recommended by the battery manufacturer. If the machine is not used for lengthy periods, the battery should be given a refresher charge every six weeks or so, to keep it in good condition.

11 Rectifier: general description

1 It is the function of the rectifier to convert the ac energy from the flywheel magneto generator into dc so that the battery can be kept charged. If the battery fails to hold its charge and is known to be in good condition, it is probable that the rectifier has failed, assuming there is output from the generator.
2 Special test equipment is necessary to verify whether the rectifier is in working order, hence it is necessary to consult either a Bultaco agent or an auto electrical expert if the rectifier is suspect.
3 The rectifier is easily damaged and for this reason it is usually located out of harms way. Under normal conditions, this component is unlikely to cause trouble. It is, of course, found only on machines fitted with a battery.

12 Ballast resistor: function

1 On some machines, irrespective of whether the lighting system is of the direct or battery operated type, a ballast resistor is fitted to eliminate the result of any surges that many occur in the electrical system at high rpm and so give rise to overload problems. If, for example, one of the light bulbs should burn out, the ballast resistor will negate the risk of the remainder of the bulbs from following suit, as a result of the decrease in electrical load that results.
2 In some cases, failure of the stop lamp bulb may cause the engine to misfire at low speeds when the brake pedal is depressed. This is because there is no longer an adequate return path for the low tension circuit of the ignition system.
3 The ballast resistor is located where it will have a constant flow of air - an important consideration since the resistor dispenses with surplus electrical energy in the form of heat.

56

Chapter 3: Ignition and electrical systems

Fig. 3.3 Battery operated lighting system (Femastronic)

3 Main wiring harness
16 Tail lamp bracket assembly complete
19 Headlamp bulb
21 Tail lamp bulb
25 Connector
26 Horn
31 Stop lamp switch
33 Stop lamp switch wire
34 Stop lamp switch bracket
36 Wire retainer
38 Horn bracket
39 Tail lamp reinforcement
40 Tail lamp reinforcement
42 Number plate bracket
43 Rectifier
44 Battery
45 Switch assembly
46 Battery carrier
47 Battery cover
48 Filler cap - 3 off
49 Filler cap washer - 3 off
50 Stop lamp switch protector
51 Front brake stop lamp switch
52 Front brake stop lamp switch bracket
53 Rectifier bracket
54 Switch protector
55 Tail lamp bracket gasket
56 Tail lamp bracket gasket
57 Reflector bush - 2 off
60 Silentbloc battery mounting - 2 off
63 Bolt - 2 off
64 Nut - 5 off
65 Nut
66 Nut 2 or 3 off (depending on model)
68 Flat washer - 4 off
69 Flat washer - 2 off
71 Screw
72 Screw - 2 off
73 Screw
74 Allen screw
75 Machine screw - 2 off
76 Star washer - 4 off
78 Star washer - 4 off
80 Nylon locknut
81 Cable strap - 6 off
82 Strap
84 Nut

13 Headlamp: replacing bulbs and adjusting beam height

1 To remove the headlamp rim, slacken the screw at the top of the headlamp shell. The rim will then pull off, complete with the reflector unit and bulb(s).

2 The reflector unit contains a double-filament bulb that provides the main and dipped headlamp beams. The double-filament headlamp bulb is controlled from a dipswitch mounted on the handlebars. Some models fitted with battery operated lighting may have an extra parking bulb mounted in the headlamp reflector.

3 It is not necessary to refocus the headlamp when a new main bulb is fitted because the bulbs are of the prefocus type. To release the bulb holder from the reflector, twist and pull.

4 Beam height is adjusted by slackening the two headlamp shell retaining bolts and tilting the headlamp either upwards or downwards. Adjustments should always be made with the rider normally seated.

5 UK lighting regulations stipulate that the lighting system must be arranged so that the light will not dazzle a person standing in the same horizontal plane as the vehicle at a distance greater than 25 yards from the lamp, whose eye level is not less than 3 feet 6 inches above that plane. It is easy to approximate this setting by placing the machine 25 yards away from a wall, on a level road, and setting the beam height so that it is concentrated at the same height as the distance from the centre of the headlamp to the ground. The rider must be seated normally during this operation.

14 Stop and tail lamp bulb: removal and replacement

1 The tail and stop lamp assembly has a single bulb, fitted with twin filaments. The base of the bulb has offset pins, to prevent it from being reversed in the bulbholder.

2 To change the bulb, rotate the lens in a clockwise direction and pull it from the assembly. The bulb can now be released from its bayonet fitting.

15 Stop lamp switch: adjustment

1 A stop lamp switch is fitted to the right-hand frame tube, close to the rear brake pedal on all models except the Sherpa. In the latter case it is operated by the brake rod and is attached to a plate welded on to the left-hand swinging fork tube. It is in the off position when the plunger is depressed and is held in this position when the pedal is at rest by means of a metal strip attached to the brake pedal. The reverse applies to the Sherpa models, where the switch is in the off position until the plunger is extended.

2 The stop lamp should operate after the brake pedal has moved about ¼ inch. To adjust the operating time, move the switch either backwards or forwards in the lug; the switch body is threaded for adjustment purposes.

16 Speedometer lamp: replacement

1 The speedometer head supplied with machines that have lighting system fitted as standard has a dial illuminating bulb fitted. The bulb and bulbholder are found on the underside of the instrument case, making renewal of the bulb easy.

17 Horn: alternator types

1 An electric horn is included in the specification of all machines fitted with a lighting system. Because an audible warning of approach is a legal requirement, an alternative bulb

horn must be fitted if the machine has no lighting system but is used on the road.

18 Wiring: layout and inspection

1 The wiring harness is colour-coded and will correspond with the accompanying wiring diagrams.
2 Visual inspection will show whether any breaks or frayed outer coverings are giving rise to short circuits. Another source of trouble may be the connectors, particularly where the connector has now been pushed home fully in the outer casing.
3 Intermittent short circuits can often be traced to a chafed wire that passes through or close to a metal component, such as a frame member. Avoid tight bends in the wire or situations where the wire can become trapped or stretched between casings.

19 Lighting and handlebar switches: general

1 Machines fitted with a battery operated lighting system have a separate lighting switch, actuated by a key. The switch is not repairable and if it malfunctions, it must be renewed.
2 Remember that if the switch is renewed, a new key must be provided to match the new switch.
3 Machines fitted with a direct lighting system have the headlamp switch mounted on the left-hand end of the handlebars. Other handlebar operated switches include a dipswitch, for dipping the main headlamp beam and a 'kill' button, to isolate the ignition circuit. The latter is wired into the low tension circuit of the flywheel magneto so that it will earth the current when the switch is operated. If the ignition circuit fails, the 'kill' button and its circuit should be checked for unsuspected earthing. A separate push button operates the electric horn, where fitted.
4 On no account oil any switch or the oil will spread across the internal contacts and form an effective insulator.

20 Fault diagnosis: ignition system

Symptom	Cause	Remedy
Engine will not start	No spark at plug	Try replacement plug if gap correct. Check whether contact breaker points are opening and closing, also whether they are clean. Check whether points arc when separated. If so, replace condenser. Check ignition coil.
Engine starts but runs erratically	Intermittent or weak spark	Try replacement plug. Check whether points are arcing. If so, replace condenser. Check accuracy of ignition timing. Low output from flywheel magneto generator or imminent breakdown of ignition coil. Plug has whiskered. Fit replacement; Plug lead insulation breaking down. Check for breaks in outer covering, particularly near frame.

21 Fault diagnosis: lighting systems

Symptom	Cause	Remedy
Complete electrical failure	Faulty main switch Isolated battery	Renew switch is cause not dirty contacts. Check battery connections, also whether connections show signs of corrosion.
Dim lights, horn inoperative	Discharged battery	Recharge battery with battery charger and check whether generator is giving correct output (electrical specialist). Check rectifier.
Constantly 'blowing' bulbs	Vibration, poor earth connection	Check whether bulb holders are secured correctly. Check earth return or connections to frame.
	Faulty ballast resistor (if fitted)	Check connections to resistor. Renew resistor, if faulty.

Chapter 4 Frame and forks

Contents

General description ... 1	Rear suspension units: examination ... 11
Front forks: removal from frame ... 2	Centre and prop stands: examination ... 12
Front forks: dismantling the steering head assembly ... 3	Footrests: examination and renovation ... 13
Front forks: dismantling ... 4	Rear brake pedal: examination and renovation ... 14
Front forks: general examination ... 5	Speedometer: removal and replacement ... 15
Front forks: examination and replacement of oil seals ... 6	Speedometer drive cable: examination ... 16
Steering head bearings: examination and replacement ... 7	Speedometer drive gearbox: examination ... 17
Front forks: reassembly ... 8	Seat: removal and replacement ... 18
Frame assembly: examination and renovation ... 9	Fault diagnosis: frame and forks ... 19
Swinging arm rear suspension ... 10	

Specifications

Front forks

	Alpina	Frontera	Pursang	Sherpa
Bultaco models				
Oil content per leg	175cc	230cc	230cc	180cc
Oil viscosity	SAE 30	SAE 30	SAE 30	SAE 10

Steering head assembly

	Alpina	Frontera	Pursang	Sherpa
Diameter of ball bearings in.	3/16	3/16	–	3/16
Arrangement	Loose	Loose	Caged*	Loose
Number of ball bearings/race	22	22	–	22

*SKF 30.205

Rear suspension

	Alpina	Frontera	Pursang	Sherpa
Range of adjustment	5 way	5 way	5 way	3 way
Travel inches ...	3 15/16	6.3	6.41	3 15/16

1 General description

The Bultaco models covered by this manual can be divided into two categories as far as the frame design is concerned. The Frontera and Pursang models have a loop type frame with swinging arm rear suspension, in which duplex tubes branch out from the lower end of the front down tube to form a cradle for the engine/gear unit before sweeping upwards at the rear to unite in a loop at the rear of the seat. The Alpina and Sherpa models employ a somewhat similar layout, but have a vertical seat tube immediately to the rear of the engine/gear unit in place of the angled tubes that branch out from the end of the top frame tubes, one on either side of the frame.

Front forks of the hydraulically damped slim-line variety are fitted to all models. The Alpina and Sherpa models are distinctive in having a different lower fork leg casting, in which the wheel spindle is carried in a forward position. The forks fitted to the Frontera and Pursang models are designed to have an extra inch of travel - 7½ inches against 6½ inches of the other machines.

2 Front forks: removal from frame

1 It is unlikely that the front forks will need to be removed from the frame as a complete unit, unless the steering head bearings require attention or the forks are damaged in an accident.

2 Commence operations by resting the machine on a stout wooden box placed under the crankcase, so that both wheels are clear of the ground. The machine must stand firmly in this position and be on level ground. On the latest Frontera models, a folding centre stand is provided, which can be used to good effect. Disconnect the front brake cable from the brake operating arm and from the cable adjuster on the front brake plate. Disconnect the speedometer drive cable from the brake plate by unscrewing the circular gland nut (Alpina and Frontera models only).

3 Loosen the two nuts and bolts that retain the torque arm to the brake plate and the front forks and detach the arm from the brake plate. Slacken the four spindle clamp nuts (two on each side - Alpina and Sherpa models) or the clamp nut and bolt on

2.2 Disconnect the front brake cable from the operating arm

2.2a Remove adjuster from front brake plate

2.3 Slacken the spindle clamp bolts

2.3a Withdraw spindle after removing end nut

2.3b Don't misplace spacer

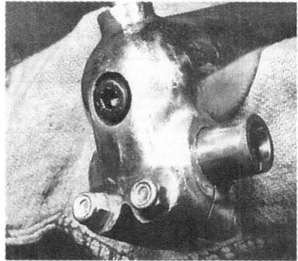
2.3c No need to displace split bushes from fork ends

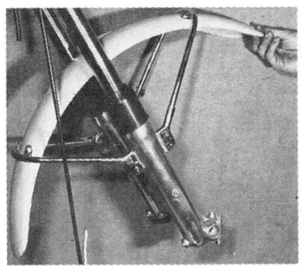
2.4 Remove front mudguard complete with stays

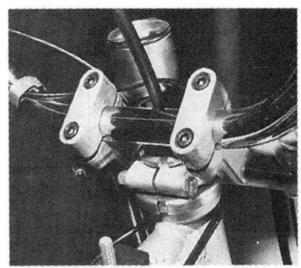
2.6 Split clamps retain handlebars on Sherpa models

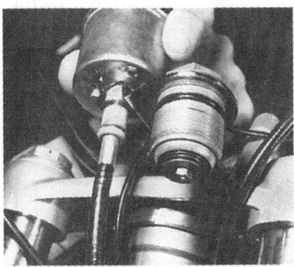
2.7 Remove filler bolt from top of each fork leg

2.7a Slacken the upper ...

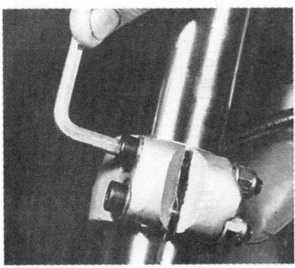
2.7b ...and lower fork yoke clamps

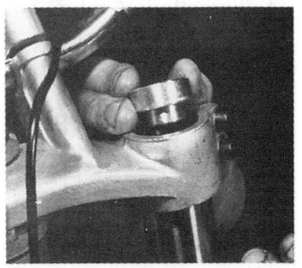
2.7c Some forks have shouldered spacers at top of each fork leg

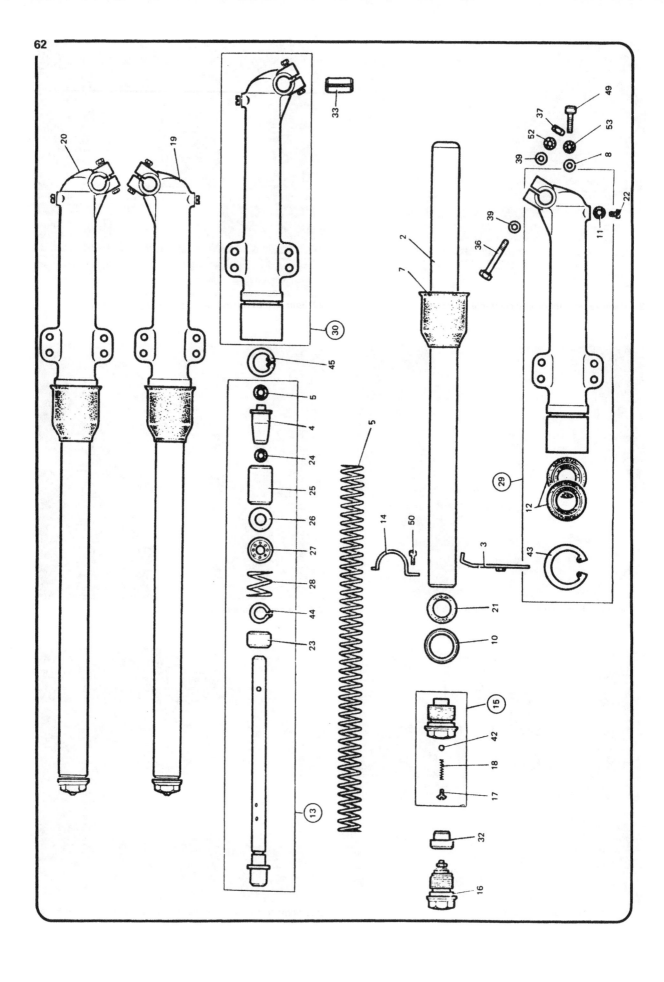

Chapter 4: Frame and forks

each side (Frontera and Pursang models). Remove the spindle nut and withdraw the wheel spindle from the left-hand side of the machine. The front wheel can now be withdrawn from the fork legs, complete with the front brake plate. Note that there is no necessity to withdraw the split bushes from either of the fork leg clamps. Do not misplace the spacer that will fall free as the spindle is withdrawn.

4 Remove the front mudguard, which is held by short stays to either the lower fork yoke (Alpina, Frontera and Pursang models) or to lugs on the lower fork leg (Sherpa models). There is no necessity to detach the stays from the mudguard itself unless they require attention.

5 On machines fitted with a lighting system, remove the front of the headlamp and detach the electrical connections, making notes to aid their eventual replacement in the correct order. Detach the headlamp by removing the two bolts that retain the shell to the headlamp brackets around the fork legs. Unscrew the speedometer drive cable from the speedometer head, if the latter is fitted in the vicinity of the upper fork yoke. It is retained by a circular gland nut.

6 Detach the various control cables from the handlebar levers, or remove the levers complete with cables. This also applies to any electrical switches, which are best removed with cables attached. Take off the handlebars, which are retained to the upper fork yoke by split clamps using either bolts or Allen screws, two on each side.

7 Remove the filler bolts that thread into the top of each fork leg and drain off the damping oil by unscrewing the drain plug at the base of each leg. If a speedometer is fitted above the steering head, the left-hand filler bolt will hold the mounting bracket captive. Lift out the shouldered spacers below, if fitted. Slacken the bolts or Allen screws in the upper and lower fork yokes and pull the fork legs downwards, one at a time, to release them from the fork yokes as complete units. If they are difficult to move, the joint in the fork yokes can be spread with a stout screwdriver, to ease their passage. Place the fork legs aside for individual dismantling.

3 Front forks: dismantling the steering head assembly

1 To dismantle the steering head assembly, unscrew and remove the nut at the top of the steering head stem. Slacken the pinch bolt through the rear portion of the upper fork yoke and pull the yoke upwards from the steering head stem.

2 To release the lower fork yoke and separate the steering head bearing assembly, unscrew the slotted nut below the yoke that has just been removed. When this nut has been unscrewed completely, the steering head stem, complete with lower fork yoke, can be withdrawn from the bottom of the steering head.

3 Bearing arrangements differ. Some machines have loose ball bearings, which will be displaced immediately the cups and cones commence to separate. Take precautions to catch these bearings by wrapping rag around the joints, so that none is lost. Where caged bearings are fitted, no such precaution is necessary.

4 On machines fitted with a steering damper, it will be necessary to remove this assembly first, even before the handlebars are detached. The damper knob and rod will unscrew from the lower damper plate assembly and can be withdrawn upwards. The damper plate assembly is held by a single screw on the underside of the lower fork yoke, after the rod has been withdrawn. If the screw is removed, the assembly is free to be taken off.

4 Front forks: dismantling

1 Although there are minor differences in the construction of the forks fitted to the various models covered by this manual, the same basic dismantling and reassembly procedure applies to all. Strip and rebuild each fork leg in turn, so that there is no possibility of inadvertently mixing the component parts.

2 Remove the Allen screw recessed into the base of the fork

Fig. 4.1 Front forks (all models except Pursang)

2 Stanchion - 2 off
3 Headlamp support - 2 off
4 Base of damper unit - 2 off
5 'O' ring - 2 off
6 Fork spring - 2 off
7 Dust cover - 2 off
8 Washer - 2 off
11 Drain screw washer - 2 off
12 Oil seal - 4 off
13 Damper assembly complete - 2 off
14 Headlamp support clamp - 2 off
15 Stanchion plug assembly complete - 2 off
16 Stanchion plug assembly - alternative type (2 off)
17 Stanchion plug screw - 2 off
18 Valve spring - 2 off
19 Right-hand fork leg
20 Left-hand fork leg
21 'O' ring - 2 off
22 Drain screw - 2 off
23 Damper piston - 2 off
24 Damper base washer - 2 off
25 Damper guide - 2 off
26 Damper washer - 2 off
27 Distributor ring - 2 off
28 Damper spring - 2 off
30 Lower fork leg - 2 off
32 Spacer - 2 off
33 Spacer for wheel spindle, left-hand side only
36 Bolt - 4 off
37 Nut - 4 off
39 Flat washer - 2 off
42 Ball bearing - 2 off
43 Circlip - 2 off
44 Circlip
45 Circlip - 2 off
49 Allen screw - 2 off
50 Allen screw - 2 off
52 Star washer
53 Star washer

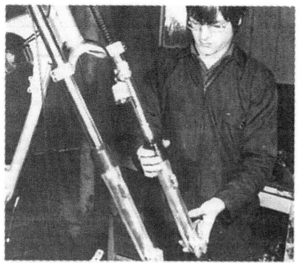

2.7d Pull fork legs downwards to release them from yokes

2.7e Drain plug at base of each leg facilitates oil drain off

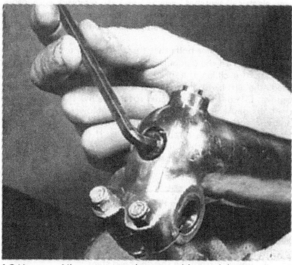

4.2 Unscrew Allen screw to release stanchion and damper assembly

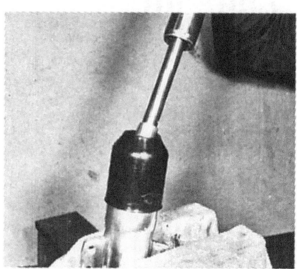

4.2a Pull stanchion from lower fork leg and ...

4.2b ...take off the dust cover

4.2c Oils seals lie inside lower fork leg

Chapter 4: Frame and forks

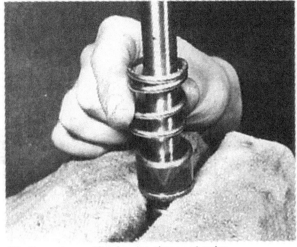

4.3 Compression spring is contained within each stanchion

4.3a Damper assembly, showing piston and spring

4.3b Remove circlip at lower end of damper rod ...

4.3c ... to release damper base washer and guide bush

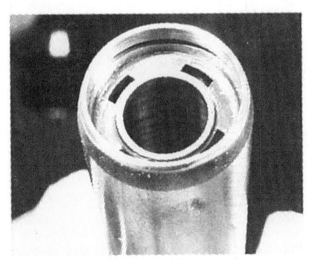

4.3d The restrictor within the stanchion

leg. When this is free, the lower leg can be pulled off the fork stanchion. Lift the dust cover from the top of the lower fork leg and remove the large diameter circlip within the fork leg housing. The two oil seals will be found immediately below the circlip, which can now be prised out of position.

3 If the compression spring is withdrawn from within the stanchion, the main portion of the damper assembly can be shaken out. It comprises the damper rod, piston, circlip and damper spring. The remainder of the damper parts are restrained from coming out by the restrictor, which is fixed with the stanchion and acts as a stop. The damper washer, damper guide bush and damper base washer will be retained within, but can be released from the other end of the stanchion by removing the circlip at the extreme end. The fork leg is now dismantled completely. Repeat, if necessary, for the other fork leg.

5 Front forks: general examination

1 Apart from the oil seals and wear that may have occurred between the sliding parts, it is unlikely that the forks will require any additional attention unless the fork springs have become compressed or the forks have become bent in an accident. The free length of the springs can be compared with that of new

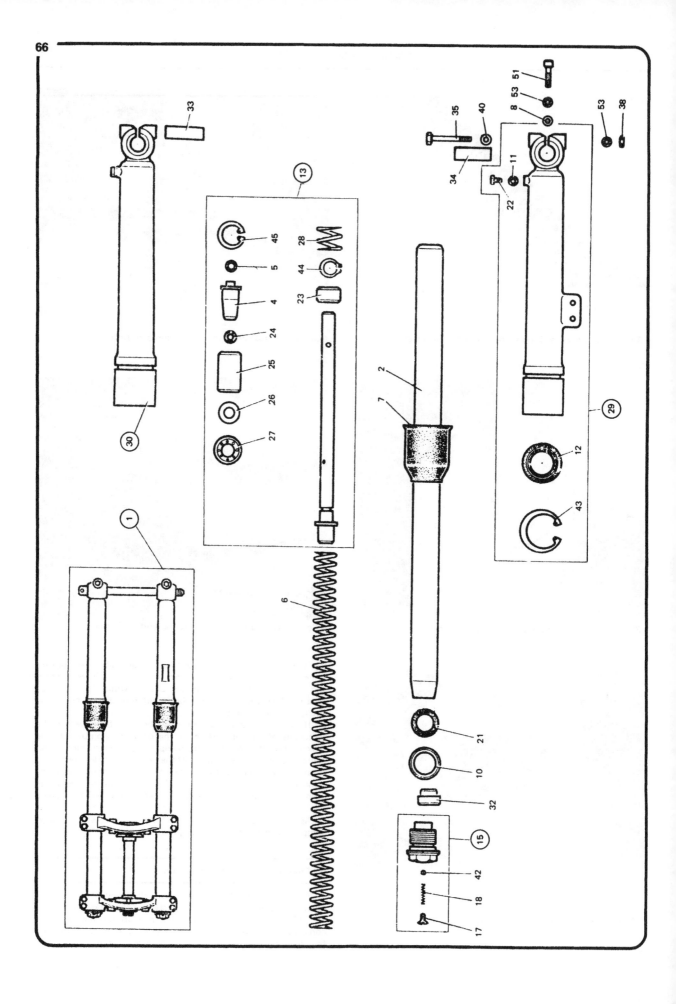

Chapter 4: Frame and forks

ones. If both springs have compressed, or one more than the other, they must be renewed as a matched pair, never singly.

2 Visual examination will show whether the fork stanchions are bent or the yokes distorted. The former can be checked for straightness by rolling them on a flat surface. Do not omit to examine the lower fork legs for cracks or signs of other damage. In the majority of cases it is advisable to renew any bent or distorted parts without having them straightened. It is rarely possible to effect a satisfactory repair and even then there is no means of telling whether fatigue failure is imminent.

3 Wear between the sliding parts of the fork will cause the forks to judder when the front brake is applied and will prove a satisfactory reason for rejecting the machine in a close examination, such as during the MOT test. Since the forks do not have renewable bushes, the parts concerned will have to be renewed to effect a satisfactory repair.

4 If the fork damping has failed, giving the forks a very lively action, the most probable cause is a worn damper piston and possibly a worn inner bore of the stanchion in which it slides. Here again, all the worn parts will have to be renewed. Wear of this nature is likely to occur only after a very long period of service, or if the forks have been run with a very low oil content.

6 Front forks: examination and replacement of oil seals

1 Failure of the oil seals is usually accompanied by oil mist, then a dribble of oil down the fork leg, which gradually gets worse as the rate of wear increases. As mentioned previously, the oil seals are paired together and are quite easily removed by prising them from the top of the lower fork leg, after the dust cover and retaining circlip have been removed.

2 Replace the new seals with their lips towards the lower end of the fork leg. They can be driven into position lightly, using a close fitting socket spanner as a drift. Remember to lubricate the lips before the stanchion is re-inserted.

3 Make sure the dust covers are in good condition and free from cracks or splits. They perform an essential role in keeping out dust, which would otherwise greatly shorten the working life of the seals.

7 Steering head bearings: examination and replacement

1 Before reassembling the front forks, examine the steering

Fig. 4.2 Front forks (Pursang models only)

1 Front fork assembly complete
2 Stanchion - 2 off
4 Base of damper unit - 2 off
5 'O' ring - 2 off
6 Fork spring - 2 off
7 Dust cover - 2 off
8 Washer - 2 off
10 Stanchion plug washer - 2 off
11 Drain plug screw washer - 2 off
12 Oil seal - 2 off
13 Damper assembly complete - 2 off
15 Stachion plug assembly complete - 2 off
17 Stanchion plug screw - 2 off
18 Valve spring - 2 off
21 'O' ring - 2 off
22 Drain screw - 2 off
23 Damper piston - 2 off
24 Damper base awsher - 2 off
25 Damper guide - 2 off
26 Damper washer - 2 off
27 Distributor ring - 2 off
28 Damper spring - 2 off
29 Lower fork leg complete - 2 off
32 Spacer - 2 off
34 Spacer for wheel spindle, right-hand side only
35 Bolt - 2 off
38 Nut - 2 off
40 Flat washer - 2 off
43 Circlip - 2 off
44 Circlip - 2 off
45 Circlip - 2 off
51 Allen screw - 2 off
53 Star washer - 4 off

6.2 Don't forget to replace circlip that retains oil seals in lower fork leg

8.5 Do not omit to tighten the torque arm nuts and bolts

8.6 Clip restrains speedometer cable from touching front tyre

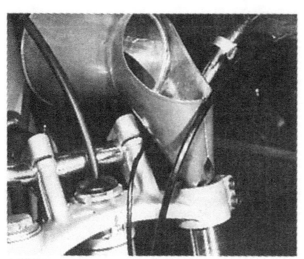

8.7 Refill forks with recommended quantity of oil

head races. If the bearings are of the cup and cone type, with loose ball bearings, check that the bearing tracks are free from cracks, flaking or indentations. If signs of wear are evident, renew the cups and cones as a set, and use new ball bearings.

2 Ball bearings are cheap. Replace the complete set without question if the originals are marked or discoloured. Use grease to hold the loose ball bearings in place whilst the steering head is reassembled. Note that there should always be space for the insertion of one extra bearing, to prevent the bearings from packing up too close and skidding on each other.

3 When caged bearings are fitted, the assembly should be washed out with petrol and allowed to dry. Any wear will be evident. If the bearing is noisy or has rough spots when rotated, renew it without question.

8 Front forks: reassembly

1 To reassemble the front forks, follow the dismantling procedure in reverse. Take particular care when passing the sliding fork members through the oil seals because the seals are very easily damaged. It is advisable to smear the sliding members with grease as well as the inside lips of each seal.

2 Tighten the steering head carefully, so that all play is eliminated without placing undue stress on the bearings. The adjustment is correct if all play is eliminated and the handlebars will swing to full lock of their own accord when given a push on one end.

3 It is possible to place several tons pressure on the steering head bearings if they are over-tightened. The usual symptom of over-tight bearings is a tendency for the machine to roll at low speeds, even though the handlebars may appear to turn quite freely.

4 If after assembly it is found that the forks are incorrectly aligned or unduly stiff in action, loosen the front wheel spindle, and two top fork leg nuts, and the pinch bolts in both the top and bottom yokes. The forks should then be pumped up and down several times to realign them. Retighten all the nuts and bolts in the same order, finishing with the steering head pinch bolt.

5 This same procedure can be adopted if the forks are misaligned after an accident. Often the legs wil twist within the fork yokes giving the impression of more serious damage, even though no structural damage has occurred.

6 Note that the damper assembly cannot be built up as a complete unit and then inserted in the stanchion, because part of it must be held captive within the stanchion. The earlier sequence of photographs outlines the manner in which reassembly is accomplished Do not omit the clip to hold the speedometer cable - Alpina and Frontera models only. The damper assembly base can be fitted after the fork spring has pushed the damper assembly through the restrictor and should then be aligned with the lower fork leg to facilitate replacement of the Allen screw. The cap on the upper end of the damper rod will be retained by the fork spring pressure alone.

7 Do not omit to refill each fork leg with the recommended quantity and grade of damping fluid:

Alpina models	175cc	SAE 30
Frontera models	230cc	SAE 30
Pursang models	210cc	SAE 30
Sherpa models	150cc	SAE 10

9 Frame assembly: examination and renovation

1 The frame should not require attention, with the possible exception of the swinging arm bearings if it has seen lengthy service, unless it has been damaged in an accident. Frame repairs and straightening are best entrusted to a frame repair specialist, who will have all the necessary jigs and mandrels essential for preserving correct alignment. Even then, no account can be made for stresses that may eventually lead to fatigue failure. If the frame is badly bent, a factory replacement is much more ad-

Chapter 4: Frame and forks

10.4 Detach only the lower end of each suspension unit

10.5 Remove pivot spindle, then ...

10.5a ... withdraw swinging arm fork complete from rear

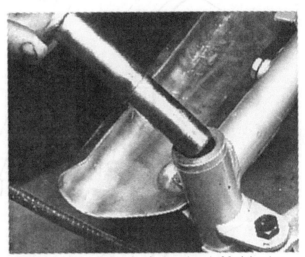
10.6 Withdraw the steel bushes from each end of fork bearing

10.8 Grease bushes thoroughly, before replacement

visable, or if the machine is used mainly for competition work, substitute such as the High-Boy frame.

2 If the machine is stripped for an overhaul, this affords an excellent opportunity to check the frame for cracks or other damage that may have occurred during service. Damage of this nature is comparatively rare, but if caught in time, a much more expensive repair at a later date can be obviated. A very careful check is essential.

10 Swinging arm rear suspension: examination and renovation

1 After an extended period of service, play will develop between the pivot pin and bushes of the swinging arm rear suspension assembly, which can be detected by pulling and pushing sideways on the rear wheel. If there is any perceptable movement that cannot be attributed to worn wheel bearings, the swinging arm assembly is in need of renovation.

2 To gain access, arrange the machine so that the rear wheel is well clear of the ground. Remove the nut from the end of the brake operating rod and withdraw the rod clear of the brake operating arm. Loosen the nuts and bolts that retain the torque arm to the rear brake plate and detach the torque arm from the brake plate itself.

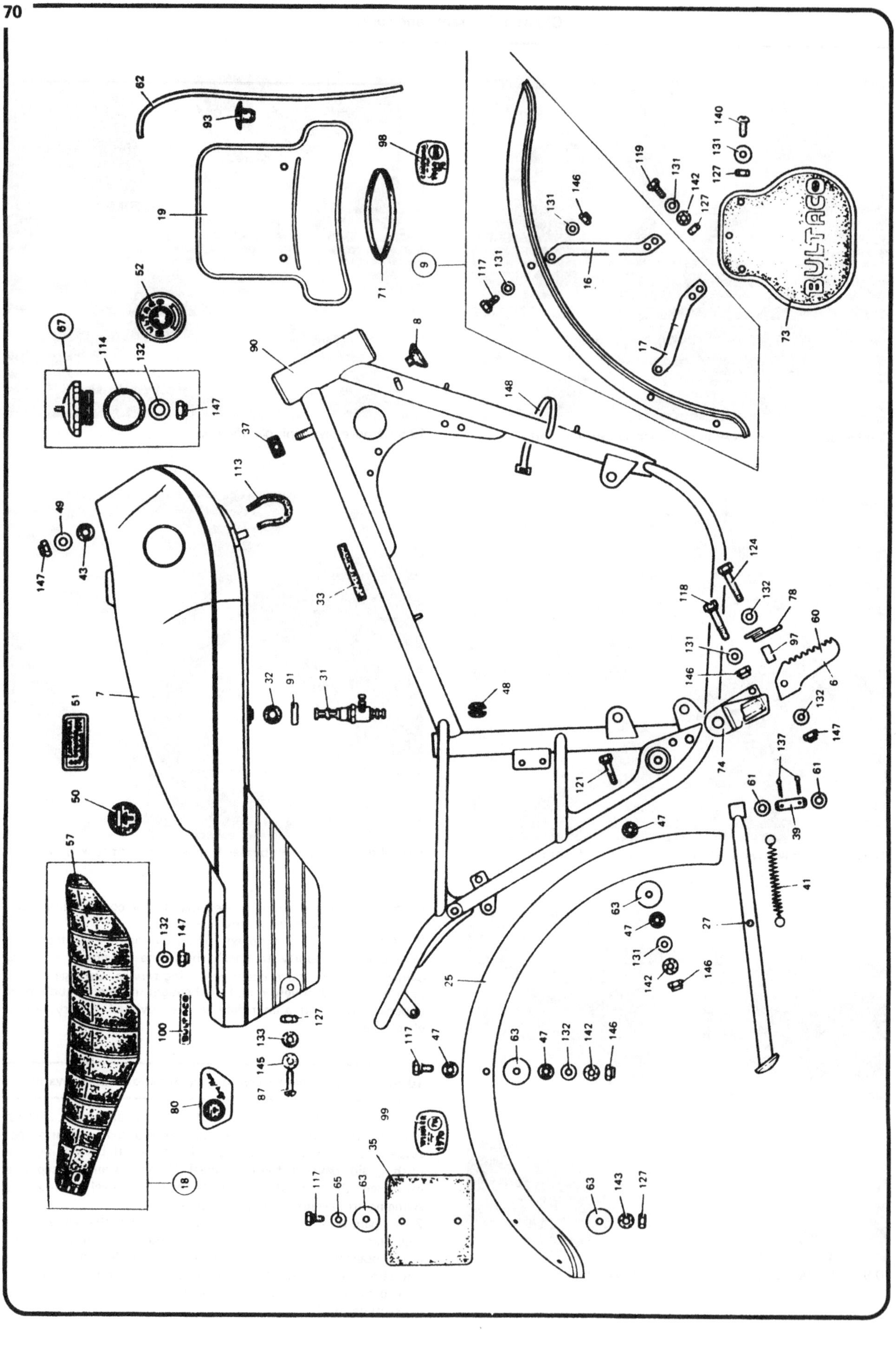
BULTACO

Chapter 4: Frame and forks

3 Detach the rear chain by withdrawing the spring link, preferably when the links either side are on the rear wheel sprocket. This makes removal easier. Slacken and remove the rear wheel spindle nut and washer and detach the speedometer drive cable from the gearbox, if the latter is of the type attached to the rear wheel (Sherpa models). Withdraw the rear wheel spindle and remove the rear wheel from the swinging arm fork, lifting the rear of the machine to gain sufficient clearance, if necessary.

4 Detach the lower connection of each rear suspension unit with the swinging arm fork. A screw with a large head is used on some models, a nut washer and bolt on others. On the Alpina and Frontera models it is also advisable to remove the chainguard, which is retained by two bolts, and on the Sherpa models, the alternative type of stop lamp switch which is bolted to a plate welded to the left-hand side of the swinging arm fork.

5 Unscrew the pivot nut on the left-hand side of the swinging arm fork assembly. Take off the washer and withdraw the pivot spindle from the right hand side of the machine. The swinging arm fork assembly can now be withdrawn from the rear of the machine.

6 The steel bushes that act as the rear suspension bearings can be withdrawn from each side of the swinging arm fork assembly without difficulty. These are the components subjected to most wear and they should be renewed if there is any play between the pivot spindle. At the same time, check the pivot spindle itself for wear and preferably renew the entire bush and shaft assembly as a complete set. It is also advisable to check the shaft for straightness if the original is to be replaced. If it is bent, renew it rather than attempt to straighten it. A good indication of whether the shaft is bent can be obtained from rolling it on a flat surface.

7 Phosphor bronze bushes are fitted to each end of the swinging arm fork and these too should be checked for wear though they are likely to give very long service before replacement is necessary. They are pressed into position and are much more difficult to remove without risk of damage. It is recommended that the removal and reinsertion of new phosphor bronze bushes is best entrusted to a Bultaco agent, who will have all the necessary service equipment.

8 When reassembling the swinging arm fork, grease all the components very thoroughly and finish off with a grease gun, continuing pumping until grease exudes from each end of the bearing. If the old grease is removed and the assembly is packed with new grease prior to reassembly, this task is made very much easier. Do not omit to replace any rubber spacers that may have been fitted originally, as they also act as dust excluders.

11 Rear suspension units: examination

1 Only a limited amount of dismantling can be undertaken because the damper unit is sealed and cannot be replaced. If the unit leaks oil, or if the damping action is lost, the suspension unit must be renewed as a whole. Always renew the units as a matched pair, never singly.

2 The suspension units are of the three or five position type and can be adjusted whilst in position on the machine, to give different spring settings. This facility is provided by a cam cut into a lower spring abutment, which engages with a peg in the lower leg of the suspension unit. Rotating the spring abutment will cause it to either rise or fall, varying the spring tension accordingly. A special peg tool is available for this purpose. It follows that both suspension units should always have an identical setting.

12 Centre and prop stands: examination

1 A prop or side stand is provided on most models, attached to the right-hand swinging arm fork leg. The stand should be inspected periodically to ensure it is securely attached to the machine and that the extension return spring has not weakened.

Fig. 4.3 Frame assembly, Sherpa T models

6 Footrest, left-hand side
7 Petrol tank
8 Rubber rest (petrol tank) - 2 off
9 Front mudguard assembly complete
16 Front left-hand mudguard stay
17 Rear left-hand mudguard stay
18 Seat assembly complete
19 Front number plate
25 Rear mudguard
27 Prop stand
31 Petrol tap
32 Petrol tap washer
33 Rubber rest (petrol tank)
35 Rear number plate
37 Rubber rest (petrol tank)
39 Prop stand spindle
41 Prop stand return spring
43 Rubber washer
47 Rubber washer - 4 off
48 Grommet
49 Petrol tank washer
50 Sherpa T transfer - 2 off
51 Championship transfer
52 Petrol tank transfer - 2 off
57 Seat cover
60 Footrest, right-hand side
61 Prop stand spindle washer - 2 off
62 Petrol tank cap breather tube
63 Washer - 2 off
65 Washer - 2 off
67 Petrol tank cap assembly
71 Rubber band
73 Front wheel mudflap
74 Footrest bracket - 2 off
78 Footrest return spring - 2 off
80 Sammy Miller replica transfer
87 Side panel screw - 2 off
90 Frame assembly complete
91 Dished washer
93 Breather plug
97 Footrest bush - 2 off
98 Spanish Championship Transfer
99 European Trials Champion transfer
100 Bultaco transfer
113 Equalising tube
114 Petrol tank cap gasket
117 Bolt - 6 off
118 Bolt - 3 off
119 Bolt - 6 off
124 Bolt - 2 off
127 Nut - 5 off
131 Flat washer - 30
132 Flat washer - 4 off
133 Rubber washer - 2 off
137 Split pin - 2 off
140 Round head screw - 4 off
142 Star washer
143 Star washer - 2 off
146 Nylon locknut - 15 off
147 Nylon locknut - 6 off

72

Chapter 4: Frame and forks

If the stand happens to extend or drop whilst the machine is in motion, it may cause a serious accident.

2 A centre stand is fitted to the Frontera models and similar advice applies. This stand is subject to greater risk of damage, as it is located on the underside of the machine.

13 Footrests: examination and renovation

1 Footrests of the spring-loaded folding type are fitted to all models. The risk of damage is greatly reduced by their folding action, but in the event of a footrest becoming bent, it must be removed from the frame for straightening.

2 To straighten a damaged footrest, clamp it in a vice and heat with a blowlamp until the area affected is at dull red heat. Use a metal pipe pushed over the footrest to provide the necessary leverage for straightening, or if this is not convenient, resort to the careful use of a hammer. Never try to straighten a footrest without heating or there is risk of it snapping off.

14 Rear brake pedal: examination and information

1 Much the same advice applies as given in the preceding section for footrests. The pedal must be detached from the machine and heat must be applied whilst straightening takes place.

2 If the brake pedal is badly deformed, it is preferable to fit a new replacement. A straightened pedal can be seriously weakened without this being obvious and may fracture without warning, most probably when it is needed most!

15 Speedometer: removal and replacement

1 The speedometer head is attached to a bracket held by the left-hand top filler bolt of the telescopic forks on most models, or in the case of the Sherpa models, by a small bracket close to the right-hand crankcase cover. The former has a rubber mounting to isolate the speedometer head from shocks.

2 To remove the speedometer head in either case, unscrew the single mounting bolt, leaving the bracket in position, after detaching the drive cable. The latter is retained by a circular gland nut.

3 Apart from defects in the speedometer drive cable or gearbox, a speedometer that malfunctions is difficult to repair. Fit a

Fig. 4.4 Frame assembly, Pursang model

1 Frame
6 Footrest, left-hand side
7 Petrol tank
10 Front mudguard
12 Front mudguard support
18 Seat assembly complete
19 Racing number plate
25 Rear mudguard
31 Petrol tap
32 Petrol tap washer
34 Rubber rest (petrol tank) - 2 off
36 Rubber rest (petrol tank)
43 Rubber washer
49 Tank washer
50 Pursang transfer
52 Petrol tank transfer - 2 off
55 Right-hand side panel
56 Left-hand side panel
57 Seat cover
60 Footrest, right-hand side
62 Petrol tank cap breather tube
67 Petrol tank cap assembly complete
78 Footrest return spring - 2 off
87 Side panel screw - 6 off
91 Dished washer - 2 off
92 Washer
93 Breather plug
95 Engine capacity decal
97 Footrest bush - 2 off
102 Front mudguard support screw
112 Washer - 7 off
114 Petrol tank cap gasket
119 Bolt - 7 off
124 Bolt - 2 off
131 Flat washer - 9 off
132 Flat washer - 7 off
133 Rubber washer - 6 off
135 Lock washer
146 Nylon locknut - 7 off
147 Nylon locknut - 7 off
149 Grommet

15.4 Sherpa models have the small diameter speedometer head

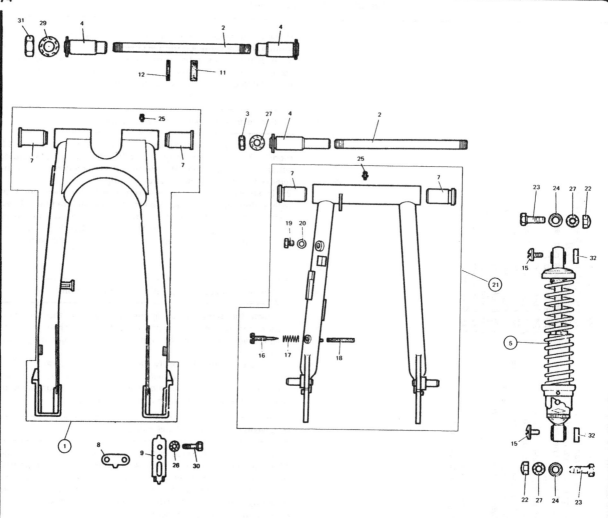

Fig. 4.5 Swinging arm rear suspension

1 Swinging arm fork assembly complete
2 Swinging arm pivot spindle
3 Swinging arm spindle retaining nut
4 Steel bush - 2 off
5 Rear suspension unit complete - 2 off
7 Bronze bush - 2 off
8 Spindle retaining plate
9 Plate
11 Rubber spacer - wide 2 off
12 Rubber spacer - narrow 2 off
13 Rear suspension unit repair kit
15 Rear suspension unit screw - 4 off
16 Chain oil adjusting screw
17 Spring
18 Oil tube
19 Oil reservoir filler cap
20 Filler cap gasket
21 Swing arm fork assembly complete - Sherpa and Alpina models only
22 Nut - 2 or 4 off, depending on model
23 Bolt - 2 or 4 off, depending on model
24 Flat washer - 2 or 4 off, depending on model
25 Grease nipple
26 Star washer - 2 off
27 Star washer - 2 or 4 off, depending on model
29 Star washer - 2 off
30 Bolt - 2 off
31 Nut (narrow) - 2 off
32 Flat washer - 4 off

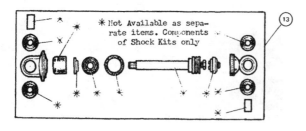

Chapter 4: Frame and forks

new replacement, or alternatively entrust the repair to an instrument repair specialist. Remember that a speedometer that functions in an efficient manner is a statutory requirement.

4 A speedometer is not fitted to the Pursang models, which are intended solely for off-road use. Only the Alpina and Frontera models have the larger diameter speedometer head equipped with a resettable odometer.

16 Speedometer drive cable: examination

1 The speedometer drive cable comprises an inner and an outer cable which can be separated. If the inner cable breaks there is no necessity to renew the complete cable assembly unless the outer cover is damaged.

2 Most cable breakages are caused by tight bends in the cable run, or damage to the outer covering which has either compressed it, or has permitted water to enter. If the mileage readings cease at the same time as the speedometer dial readings, this is a sure sign the cable is at fault, or the drive itself.

3 Do not grease the last six inches of the cable at the speedometer head end. If this precaution is ignored, grease may work into the speedometer head and immobilise the movement.

17 Speedometer drive gearbox: examination

1 On the Sherpa models, the speedometer drive is taken from a drive gearbox through which the rear wheel spindle passes. It takes up the drive from the hub of the rear wheel. Provided the gearbox is greased regularly, it should not give trouble.

2 The Alpina and Frontera models take the speedometer drive from a gearbox incorporated in the hub of the front wheel. Here again, regular lubrication should ensure no troubles are encountered.

3 On Sherpa models, the speedometer gearbox can be pulled off the rear hub after the wheel spindle has been withdrawn. On the Alpina and Frontera models, the drive assembly is built into the front brake plate. If the rear spindle type malfunctions, the complete unit will have to be renewed - the gearbox is not repairable. The individual drive pinions of the front brake plate arrangement can be renewed if they become damaged since the assembly is easy to dismantle and reassemble.

18 Seat: removal and replacement

1 On the Alpina models, the seat forms part of the tank fairing assembly and is attached to the rear mudguard by two small brackets at the rear end, one on each side. On the Frontera and Pursang models, it is attached by means of the two uppermost screws that pass through each side cover.

2 The Sherpa models have a combined tank and seat unit, which is detached as described in Chapter 1, Sections 5.3 - 5.4.

19 Fault diagnosis: frame and forks

Symptom	Cause	Remedy
Machine veers to left or right with hands off handlebars	Incorrect wheel alignment Bent forks Twisted frame	Check and re-align. Check and replace. Check and replace.
Machine rolls at low speeds	Overtight steering head bearings	Slacken and re-test.
Machine judders when front brake is applied	Slack steering head bearings	Tighten until all play is taken up.
Machine pitches badly on uneven surfaces	Ineffective fork dampers Ineffective rear suspension units	Check oil content. Check damping action.
Fork action stiff	Fork legs out of alignment (twisted in yokes)	Slacken yoke clamps, front wheel spindle and fork top bolts. Pump forks several times, then tighten from bottom upwards.
Machine wanders. Steering imprecise, rear wheel tends to hop	Worn swinging arm pivot	Dismantle and replace bushes and pivot shaft.

Chapter 5 Wheels, brakes and tyres

Contents

General description ... 1	Rear wheel shock absorber: examination ... 9
Front wheel: examination and removal ... 2	Rear wheel sprocket: examination and removal ... 10
Front brake assembly: examination, renovation and reassembly ... 3	Rear wheel: replacement in frame ... 11
Wheel bearings: examination and renewal ... 4	Front and rear brakes: adjustment ... 12
Front wheel: reassembly and replacement ... 5	Final drive chain: examination and lubrication ... 13
Rear wheel: examination, removal and renovation ... 6	Tyres: removal and replacement ... 14
Rear brake drum: examination, renovation and reassembly ... 7	Security bolts ... 15
Rear wheel bearings: examination and renewal ... 8	Fault diagnosis: wheels, brakes and tyres ... 16

Specifications

Wheel sizes
Front ...	21 inch daimeter — all models
Rear ...	18 inch diameter

Tyre sizes

Models	Alpina	Frontera	Pursang	Sherpa
Front in.	21 x 3.00	21 x 3.00	21 x 3.00	21 x 2.75
Rear in.	18 x 4.00	18 x 4.50	18 x 4.00 (200cc) 18 x 4.50 (250/360)	18 x 4.00

Tyre pressures

	Alpina	Frontera	Pursang	Sherpa
Trials/cross country				
Front psi	5	8	—	5
Rear psi	5	6	—	5
Trail riding (both)	8	—	—	8
Road use				
Front	14	12	N/A	N/A
Rear	14	8	N/A	N/A

Brakes

	Alpina	Frontera	Pursang	Sherpa
Front mm	140 x 35	140 x 30	125 x 25	125 x 25
Rear	140 x 40	140 x 30	140 x 30	140 x 30

1 General description

Each of the machines covered by this manual is fitted with a 21 inch diameter front wheel and a rear wheel of 18 inch diameter, the usual requirements of a machine suitable for competition and general on/off road use. The large diameter front wheel helps provide the necessary high ground clearance and accurate steering and the rear wheel is suitable for a wide cross section tyre, in order to obtain better wheel grip. The Alpina, Frontera and Pursang models have a 3.00 inch cross section front tyre fitted, only the Sherpa models have a smaller cross section tyre of 2.75 inches to give the steering greater sensitivity. All models have a rear tyre of 4.00 inch cross section, with the exception of the Frontera range, where the section is increased to 4.50 inches.

Drum brakes are used on all models. The Alpina and Frontera models have a front brake of 140 mm diameter, and the Pursang and Sherpa models one of only 125 mm diameter in view of the special requirements of these models. All have a 140 mm rear brake and all brakes are of the single leading shoe type.

2 Front wheel: examination and removal

1 To remove the front wheel, follow the procedure given in Chapter 4, Section 2, paragraphs 2 - 3. Before the wheel is removed, however, it is advisable to raise the front of the machine so that the wheel is clear of the ground and can be spun. This will enable the rim alignment to be checked and any small irregularities to be corrected by tightening the spokes in

Chapter 5: Wheels, brakes and tyres

3.1 Front brake plate will lift off hub

3.1a Small distance piece within hub centre is easily lost

3.4 Linings are of the bonded-on variety

the affected area. A certain amount of experience is called for if over-correction is to be avoided.

2 If any flats are evident in the wheel rim, they will prove much more difficult to remove and in the majority of cases the wheel will have to be rebuilt on a new rim. Apart from the effect on stability, there is greater risk of damage to the tyre bead and walls if the machine is run with a deformed wheel.

3 Check for loose or broken spokes. Tapping the spokes is the best guide to tension. A loose spoke will produce a quite different sound and should be tightened by turning the nipple in an anti-clockwise direction. Always re-check for run-out by spinning the wheel again. If excessive tightening is required, it is advisable to remove the tyre and inner tube, so that the now projecting spoke ends can be filed off. They may otherwise cause punctures.

3 Front brake assembly: examination, renovation and reassembly

1 When the front wheel has been removed from the machine, the front brake plate can be lifted off the hub. Do not misplace the small distance piece in the centre of the hub assembly, through which the wheel spindle passes. It is easily mislaid.

2 Examine the condition of the brake linings. If they are wearing thin or unevenly, the brake shoes should be replaced.

3 To remove the brake shoes from the brake plate, detach the circlip from the common pivot and pull them apart whilst lifting them upwards in the form of a 'V'. When they are clear of the brake plate, the return spring can be removed and the shoes separated.

4 Boned-on brake linings are employed. If the linings are worn or badly scored, it will be necessary to renew the brake shoes as a pair. Most Bultaco dealers can offer replacement brake shoes on a service exchange basis.

5 Before replacing the brake shoes, check that the brake operating cam is working smoothly and not binding in its pivot. The cam can be removed for greasing by unscrewing the nut on the brake operating arm and drawing the arm off, so that the spindle and cam can be pushed out of the housing in the back plate.

6 Check also the inner surface of the brake drum, on which the brake shoes bear. The surface should be smooth and free from score marks or indentations, otherwise reducing braking efficiency is inevitable. Remove all traces of brake lining dust and wipe with a rag soaked in petrol to remove any traces of grease or oil.

7 To reassemble the brake shoes on the brake plate, fit the return spring and force the shoes apart, holding them in a 'V' formation. If they are now located with the brake operating cam and pivot they can usually be snapped into position by pressing downwards. Do not use excessive force, or the shoes may be distorted permanently. Do not omit to replace the circlip on the common pivot.

4 Wheel bearings: examination and renewal

1 Wheel bearings of the ball journal type are employed, which are pre-packed with grease and do not normally require attention until play develops. When play in the wheel bearings occurs, the bearings are no longer fit for further service and must be driven out of the hub and renewed.

2 Each bearing should be driven outwards, working from the centre of the hub, after first removing any seals that may precede the bearings. Note the arrangement of the seals and 'O' rings, so that they can be reassembled in the same order. There is a bearing spacer between the two bearings which should be drifted out when the second of the two bearings is removed from the hub. It is shouldered and will help displace the bearing concerned.

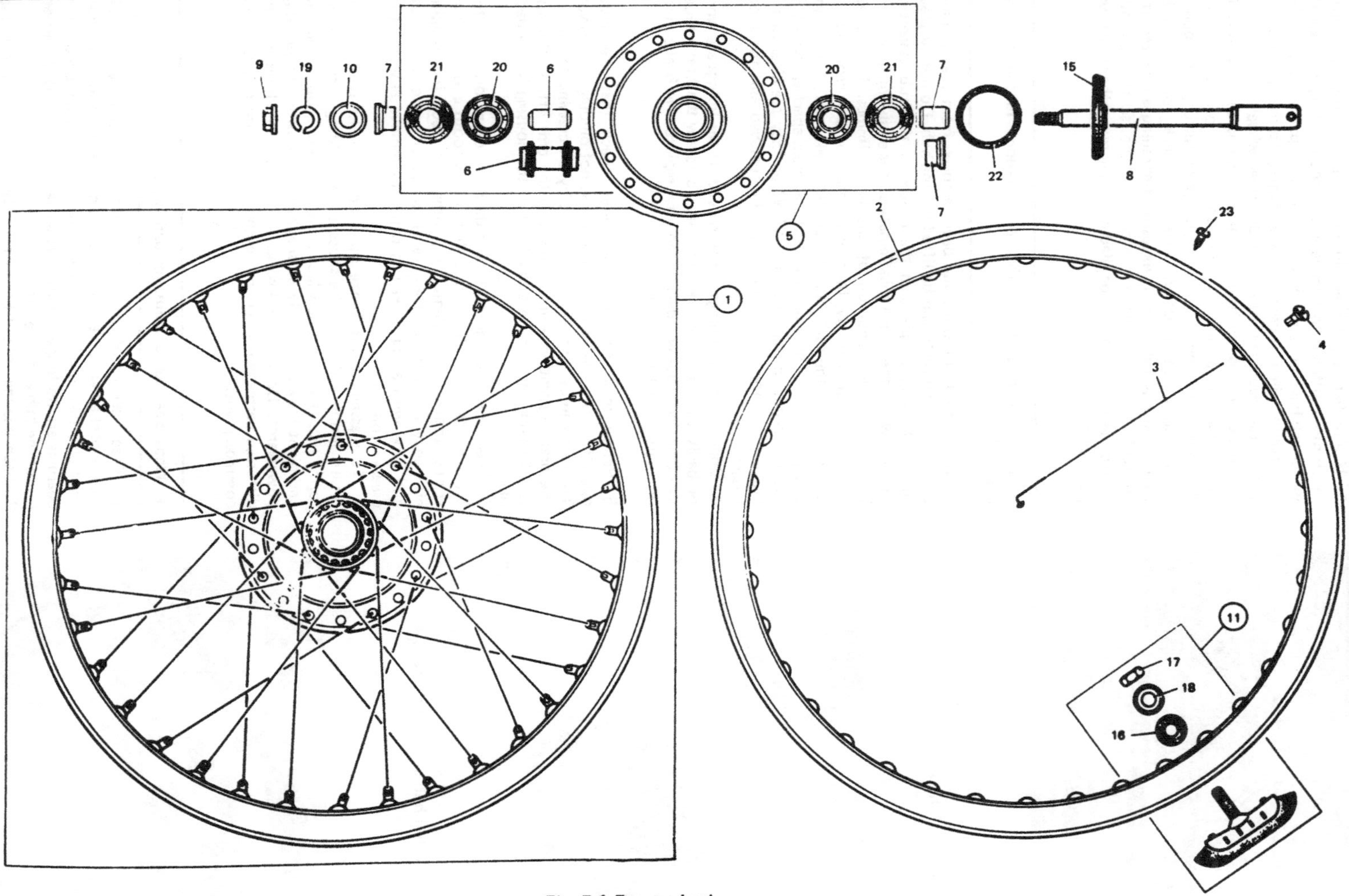

78

Fig. 5.1 Front wheel

1 Front wheel assembly complete
2 Wheel rim
3 Spoke - 36 off
4 Spoke nipple - 36 off
5 Front hub assembly complete
6 Bearing spacer
7 Bearing lock
8 Front wheel spindle
9 Wheel spindle nut
10 Spindle washer
11 Security bolt assembly
15 Hub dust protector
16 Leather washer
17 Anchor nut
18 Flat washer
19 Lockwasher
20 Wheel bearing - 2 off
21 Seat - 2 off
22 'O' ring - 2 off
23 Screw

Chapter 5: Wheels, brakes and tyres

6.2 Rear wheel will withdraw from end of swinging arm fork

7.1 Rear brake drum is retained by nine Allen screws

7.1a Rear brake is of similar design to that of front wheel

3 Remove all the old grease from the hub and bearings, giving the latter a final wash in petrol. Check the bearings for play or signs of roughness when they are turned. If there is any doubt about their condition, play safe and replace them.

4 Before replacing the bearings, first pack the hub with new grease. Then grease both bearings and drive them back into position using either a double-diameter drift or the wheel spindle.

5 If the front wheel is from an Alpina or Frontera model, it is advisable to check the condition of the front wheel speedometer drive components before the wheel is replaced in the front forks. The drive dogs form part of the front hub assembly and the drive pinions can be found within the front brake plate. Renew all worn and/or damaged parts.

6 Check also the front wheel spindle, in case it is bent.

5 Front wheel: reassembly and replacement

1 Replace the front brake plate in the brake drum and align the front wheel so that the front wheel spindle can be passed through the fork ends from the left-hand side. Do not omit the spacer on the left-hand side of the wheel hub and do not omit to replace and tighten the wheel spindle nut and the fork end clamp bolts.

2 Check that the front brake torque arm has been reconnected correctly and the nuts and bolts tightened. If this part works loose and becomes disconnected, the front brake will lock in the permanently on position, immediately it is applied. The consequences could prove fatal.

3 If the wheel has a front wheel speedometer drive, reconnect the drive cable by means of the circular gland nut. Note the front brake cable should be restrained from touching the tyre by means of the metal clip bolted to the upper of the torque arm connections.

6 Rear wheel: examination, removal and renovation

1 Before removing the rear wheel from the swinging arm fork, check for rim alignment, damage to the rim and loose or broken spokes, as described in Section 3 of this Chapter for the front wheel.

2 The rear wheel is not quickly detachable and must be withdrawn from the swinging arm fork as described in Chapter 4, Section 10, paragraphs 2 and 3.

7 Rear brake drum: examination, renovation and reassembly

1 The rear brake is of identical design to that built into the front wheel except that it is retained to the hub by nine Allen screws. This is necessary because the hub contains a series of shock absorbing rubbers with which the rear of the brake drum engages.

2 Refer to Section 3 of this Chapter for advice about the removal of the brake shoes and examination of the brake components in general. When examining the brake drum, check that the nine Allen screws in the centre are tight and that the sprocket retaining bolts around the periphery of the brake drum are also tight. Reassemble the brake shoes in the reverse order of dismantling, not forgetting to replace the circlip around their common pivot.

8 Rear wheel bearings: examination and renewal

1 The rear wheel bearings are arranged in similar fashion to those of the front wheel and have a shouldered spacer separating them. Each must be driven outwards from within the hub, using the spacer to dislodge the second bearing.

2 Check each bearing as descirbed in Section 3 of this Chapter

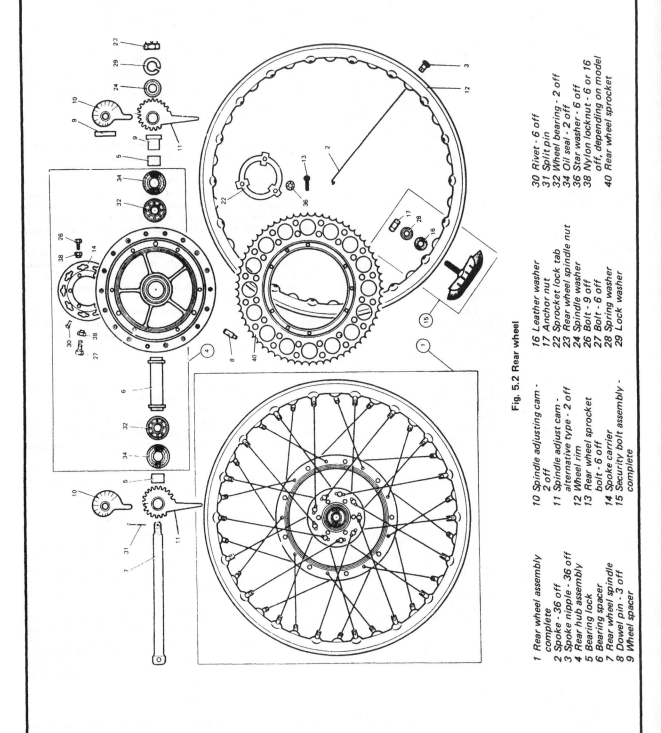

Fig. 5.2 Rear wheel

1 Rear wheel assembly - complete
2 Spoke - 36 off
3 Spoke nipple - 36 off
4 Rear hub assembly
5 Bearing lock
6 Bearing spacer
7 Rear wheel spindle
8 Dowel pin - 3 off
9 Wheel spacer
10 Spindle adjusting cam - 2 off
11 Spindle adjust cam - alternative type - 2 off
12 Wheel rim
13 Rear wheel sprocket bolt - 6 off
14 Spoke carrier
15 Security bolt assembly - complete
16 Leather washer
17 Anchor nut
22 Sprocket lock tab
23 Rear wheel spindle nut
24 Spindle washer
26 Bolt - 9 off
27 Bolt - 6 off
28 Spring washer
29 Lock washer
30 Rivet - 6 off
31 Split pin
32 Wheel bearing - 2 off
34 Oil seal - 2 off
36 Star washer - 6 off
38 Nylon locknut - 6 or 16 off, depending on model
40 Rear wheel sprocket

8.1 Wheel bearings are arranged similarly to those of front wheel

10.1 Examine sprocket teeth carefully

10.2 Sprocket is retained to brake drum by nine bolts

11.1 Snail cams make wheel alignment easy

11.1a Note return spring on brake operating arm

11.2 Torque arm nuts and bolts MUST be tightened fully

Chapter 5: Wheels, brakes and tyres

and repack the bearings and hub with new grease before reassembly. Take careful note of the arrangement of the spacers and seals, so that they are replaced in the correct order.

9 Rear wheel shock absorber: examination

1 Provision is made for the inclusion of a shock absorber in the rear hub to smooth out any surges that may otherwise cause harsh transmission. Projections on the back of the rear brake drum engage with slots in a number of synthetic rubber buffers arranged within the flanged left-hand end of the hub. They allow a certain amount of movement so that the sprocket affixed to the periphery of the brake drum can move independently of the wheel itself.
2 Excessive movement of the sprocket or traces of rubber dust are the usual indications that the shock absorber assembly requires attention. To gain access to the rubbers, unscrew the nine Allen screws in the centre of the rear brake drum and pull the brake drum cum sprocket away from the assembly. There is no necessity to disturb the wheel bearings.
3 The rubbers are easily removed and replaced with their new counterparts. If the new rubbers are more difficult to insert, a light smearing of household detergent will help. Make sure the projections on the rear of the brake drum engage correctly before the nine Allen screws are replaced and tightened fully.
4 Before suspecting the shock absorber assembly, make sure the sprocket retaining bolts have not slackened.

10 Rear wheel sprocket: examination and removal

1 Examine the teeth of the rear wheel sprocket. If they are worn, hooked, chipped or broken, the sprocket must be renewed. Damage of this nature will cause very rapid chain wear.
2 The sprocket is retained to the periphery of the rear brake drum by nine bolts which must always be kept tight. Remove them to free the sprocket. If the sprocket has to be renewed, it is advisable to renew the gearbox final drive sprocket at the same time and the rear chain. If old and new parts are run together, the overall rate of wear will be much higher as a result.
3 Always fit sprocket sizes as specified by the manufacturer. In the case of the rear wheel sprocket, the number of teeth is stamped on the face of the sprocket. The Alpina and Frontera models have a 42 tooth rear sprocket, whereas the Pursang and Sherpa models have one with 46 teeth. There are corresponding differences in the size of the gearbox final drive sprocket - see Specifications.

11 Rear wheel: replacement in frame

1 Replace the rear wheel in the swinging arm fork by reversing the procedure used for its removal. Note that wheel alignment, an important factor, is made easy by the use of the snail cam adjusters. This ensures the wheel is drawn back an equal amount on each side when checking the chain adjustment. Do not omit to replace and tension the brake operating arm return spring is disturbed.
2 Do not omit to replace and tighten the torque arm and its connections. If the torque arm works loose and becomes disconnected, the brake will be in the permanently on position immediately it is applied. This will provoke a skid that may well have fatal consequences.
3 On the Sherpa models, check that the operating arm of the speedometer drive gearbox has located with the slot in the hub flange before the rear wheel spindle is inserted and tightened. The speedometer cable should be reconnected with the circular gland nut and arranged to have an easy sweep.

12 Front and rear brakes: adjustment

1 The front brake adjuster is located on the front brake plate. The brake should be adjusted so that the wheel is free to revolve before pressure is applied to the handlebar lever and is applied fully before the handlebar lever touches the handlebar. Make sure the adjuster locknut is tight after the correct adjustment has been made.
2 The rear brake is adjusted by means of the adjusting nut on the end of the brake operating rod. Adjustment is largely a matter of personal choice, but excessive travel of the footbrake pedal should not be necessary before the brake is applied fully.
3 Efficient brakes depend on good leverage of the operating arms. The angle between the brake operating arm and the cable or rod should never exceed $90°$ when the brake is fully applied.
4 If the rear wheel has to be moved backwards to take up chain slack, or forwards when a new chain is fitted, the rear brake will require readjustment. This also applies to the stop lamp switch, when a lighting set is fitted. Refer to Chapter 3, Section 15, for adjustment of the latter.

13 Final drive chain: examination and lubrication

1 The final drive chain has its own independent lubrication by means of a controlled drip feed from an oil reservoir contained within the hollow left-hand leg of the swinging arm fork. The hexagon bolt acts as the filler cap for the reservoir and the large headed screw is the needle valve for controlling the rate of flow, which is delivered via a short tube attached to the inner portion of the fork leg. Experience will dictate the best setting; too much and oil will collect on the sidewalls of the tyre - too little and the chain will run dry. Use only a light oil - a SAE rating of 10 is correct.
2 To preserve chain longevity, it is advisable to remove the chain periodically and wash it thoroughly in a petrol/paraffin mix to remove all traces of grit, etc. When it is dry and has been brushed clean, it can be sprayed with chain lubricant now available in aerosol form, or preferably immersed in a bath of molten chain lubricant such as Linklyfe or Chainguard. In the latter case it should then be suspended lengthwise for the lubricant to solidify, before it is refitted to the machine. Aerosol lubricants can be applied with the chain in situ, but only after the chain has been brushed clean.
3 Chain adjustment is correct when there is approximately ¾ inch play in the middle of the upper run. Bultaco models have an automatic chain tensioner fitted to the lower run of the chain, which will effectively take up any reasonable amount of additional slack. It runs on a bush which should be looked at periodically for wear. Always check tension with the tensioner temporarily out of action.
4 If the chain is too slack, slacken the wheel spindle nut and spindle, also the nuts and bolts on the rear brake torque arm. Move the wheel backwards by means of the snail cam adjusters, checking to ensure the wheel has moved backwards an equal amount on each side. The cams have graduation marks to aid this. When the tension is correct tighten the spindle and spindle nut, then the torque arm nuts and bolts. Recheck the chain tension. Always check with the rear wheel in several different positions, because a chain never wears in an even fashion.
5 To check whether the chain needs replacing lay it lengthwise in a straight line and compress it, so that all play is taken up. Anchor one end and then pull on the other, to stretch the chain in the opposite direction. If the chain extends by more than the distance between the two adjacent rollers, replacement is advised.
6 When replacing the chain, make sure the spring link is positioned correctly, with the closed end facing the direction of travel. Reconnection is made easier if the ends of the chain are pressed into the teeth of the rear wheel sprocket.

13.1 Chain oiler reservoir filler cap

13.1a The adjustable needle valve meters the oil flow

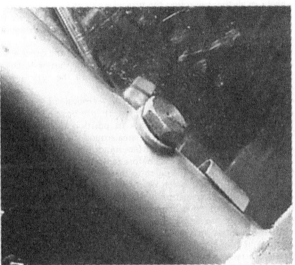
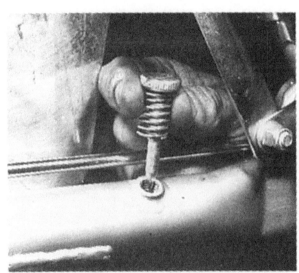

13.3 Automatic chain tensioner presses on lower chain run

13.6 Closed end of spring link must always face direction of chain travel

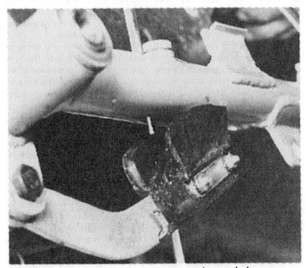
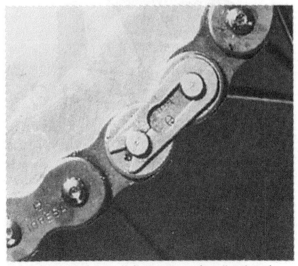

Chapter 5: Wheels, brakes and tyres

Fig. 5.3a Tyre fitting

A *Deflate inner tube and insert lever in close proximity to tyre valve*
B *Use two levers to work bead over the edge of the rim*
C *When first bead is clear, remove tyre as shown*

14 Tyres: removal and replacement

1 At some time or other the need will arise to remove and replace the tyres, either as the result of a puncture or because a replacement is required to offset wear. To the inexperienced, tyre changing represents a formidable task yet if a few simple rules are observed and the technique learned the whole operation is surprisingly simple.

2 To remove the tyre from either wheel, first detach the wheel from the machine by following the procedure in Chapters 4.2 or 5.6, depending on whether the front or the rear wheel is involved. Deflate the tyre by removing the valve insert and when it is fully deflated, push the bead of the tyre away from the wheel rim on both sides so that the bead enters the centre well of the rim. Remove the locking cap and push the tyre valve into the tyre itself.

3 Insert a tyre lever close to the valve and lever the edge of the tyre over the outside of the wheel rim. Very little force should be necessary; if resistance is encountered it is probably due to the fact that the tyre beads have not entered the well of the wheel rim all the way round the tyre.

4 Once the tyre has been edged over the wheel rim, it is easy to work around the wheel rim so that the tyre is completely free on one side. At this stage the inner tube can be removed.

5 Working from the other side of the wheel, ease the other edge of the tyre over the outside of the wheel rim that is furthest away. Continue to work around the rim until the tyre is free completely from the rim.

6 If a puncture has necessitated the removal of the tyre, re-inflate the inner tube and immerse it in a bowl of water to trace the source of the leak. Mark its position and deflate the tube. Dry the tube and clean the area around the puncture with a petrol-soaked rag. When the surface has dried, apply the rubber solution and allow this to dry before removing the backing from the patch and applying the patch to the surface.

7 It is best to use a patch of the self-vulcanising type, which will form a very permanent repair. Note that it may be necessary to remove a protective covering from the top surface of the patch, after it has sealed in position. Inner tubes made from synthetic rubber may require a special type of patch and adhesive, if a satisfactory bond is to be achieved.

8 Before replacing the tyre, check the inside to make sure the agent that caused the puncture is not trapped. Check also the outside of the tyre, particularly the tread area, to make sure nothing is trapped that may cause a further puncture.

9 If the inner tube has been patched on a number of past occasions, or if there is a tear or large hole, it is preferable to discard it and fit a replacement. Sudden deflation may cause an accident, particularly if it occurs with the front wheel.

10 To replace the tyre, inflate the inner tube sufficiently for it to assume a circular shape but only just. Then push it into the tyre so that it is enclosed completely. Lay the tyre on the wheel at an angle and insert the valve through the rim tape and the hole in the wheel rim. Attach the locking cap on the first few threads, sufficient to hold the valve captive in its correct location.

11 Starting at the point furthest from the valve, push the tyre bead over the edge of the wheel rim until it is located in the central well. Continue to work around the tyre in this fashion until the whole of one side of the tyre is on the rim. It may be necessary to use a tyre lever during the final stages.

12 Make sure there is no pull on the tyre valve and again commencing with the area furthest from the valve, ease the other bead of the tyre over the edge of the rim. Finish with the area close to the valve, pushing the valve up the tyre until the locking cap touches the rim. This will ensure the inner tube is not trapped when the last section of the beam is edged over the rim with a tyre lever.

13 Check that the inner tube is not trapped at any point. Re-inflate the inner tube, and check that the tyre is seating correctly around the wheel rim. There should be a thin rib moulded around the wall of the tyre on both sides, which should be

Fig. 5.3b Tyre fitting

D Inflate inner tube and insert in tyre
E Lay tyre on rim and feed valve through hole in rim
F Work first bead over rim, using lever in final section
G Use similar technique for second bead. Finish at tyre valve position
H Push valve and tube up into tyre when fitting final section, to avoid trapping

Security bolts (Competition models)
I Fit the security bolt very loosely when one bead of the tyre is fitted
J Then fit tyre in normal way. Tighten bolt when tyre is properly seated

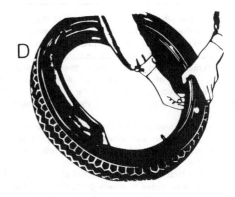

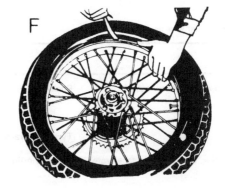

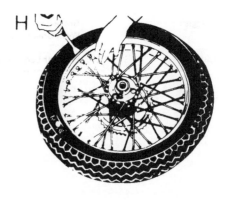
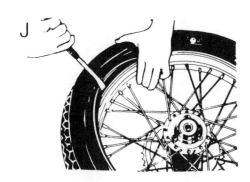
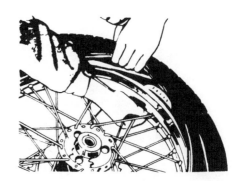

equidistant from the wheel rim at all points. If the tyre is unevenly located on the rim, try bouncing the wheel when the tyre is at the recommended pressure. It is probable that one of the beads has not pulled clear of the centre well.

14 Tyre replacement is aided by dusting the side walls, particularly in the vicinity of the beads, with a liberal coating of french chalk. Washing-up liquid can also be used to good effect, but this has the disadvantage of causing the inner surfaces of the wheel rim to rust.

15 Never replace the inner tube and tyre without the rim tape in position. If this precaution is overlooked there is good chance of the ends of the spoke nipples chafing the inner tube and causing a crop of punctures.

16 Never fit a tyre that has a damaged tread or side walls. Apart from the legal aspects, there is a very great risk of a blow-out, which can have serious consequences on any two-wheel vehicle.

17 Tyre valves rarely give trouble, but it is always advisable to check whether the valve itself is leaking before removing the tyre. Do not forget to fit the dust cap, which forms an effective second seal.

15 Security bolts

1 It is often necessary to run the tyres at low pressures, in order to obtain the benefit of greatly improved wheel grip on rough terrain. Under these circumstances, the tyre tends to creep on the wheel rim unless it can be restrained in some way. The security bolt fulfills this requirement in a simple and effective manner, to prevent the valve being torn from the inner tube body as it is dragged with the outer cover.

2 A security bolt is fitted to the rear wheel. Before attempting to remove or replace a tyre, the security bolt must be slackened off completely because it clamps the bead of the tyre to the wheel rim.

16 Fault diagnosis: wheels, brakes and tyres

Symptom	Cause	Remedy
Handlebars oscillate at low speeds	Buckle or flat in wheel rim, most probably from wheel	Check rim alignment by spinning wheel. Correct by retensioning spokes or by having wheel rebuilt on new rim.
	Tyre not straight on rim	Check tyre alignment.
Machine lacks power and accelerates	Brakes binding	Warm brake drums provide best evidence. Re-adjust brakes.
Brakes grab when applied gently	Ends of brake shoes not chamfered	Chamfer with file.
	Elliptical brake drum	Lightly skim in lathe (specialist attention needed).
Brake pull-off sluggish	Brake cam binding in housing	Free and grease.
	Weak brake shoe springs	Replace if springs not displaced.
Harsh transmission	Worn or badly adjusted chains	Adjust or replace as necessary.
	Hooked or badly worn sprockets	Replace as a pair.

Metric conversion tables

Inches	Decimals	Millimetres	Millimetres to Inches		Inches to Millimetres	
			mm	Inches	Inches	mm
1/64	0.015625	0.3969	0.01	0.00039	0.001	0.0254
1/32	0.03125	0.7937	0.02	0.00079	0.002	0.0508
3/64	0.046875	1.1906	0.03	0.00118	0.003	0.0762
1/16	0.0625	1.5875	0.04	0.00157	0.004	0.1016
5/64	0.078125	1.9844	0.05	0.00197	0.005	0.1270
3/32	0.09375	2.3812	0.06	0.00236	0.006	0.1524
7/64	0.109375	2.7781	0.07	0.00276	0.007	0.1778
1/8	0.125	3.1750	0.08	0.00315	0.008	0.2032
9/64	0.140625	3.5719	0.09	0.00354	0.009	0.2286
5/32	0.15625	3.9687	0.1	0.00394	0.01	0.254
11/64	0.171875	4.3656	0.2	0.00787	0.02	0.508
3/16	0.1875	4.7625	0.3	0.01181	0.03	0.762
13/64	0.203125	5.1594	0.4	0.01575	0.04	1.016
7/32	0.21875	5.5562	0.5	0.01969	0.05	1.270
15/64	0.234375	5.9531	0.6	0.02362	0.06	1.524
1/4	0.25	6.3500	0.7	0.02756	0.07	1.778
17/64	0.265625	6.7469	0.8	0.03150	0.08	2.032
9/32	0.28125	7.1437	0.9	0.03543	0.09	2.286
19/64	0.296875	7.5406	1	0.03947	0.1	2.54
5/16	0.3125	7.9375	2	0.07874	0.2	5.08
21/64	0.328125	8.3344	3	0.11811	0.3	7.62
11/32	0.34375	8.7312	4	0.15748	0.4	10.16
23/64	0.359375	9.1281	5	0.19685	0.5	12.70
3/8	0.375	9.5250	6	0.23622	0.6	15.24
25/64	0.390625	9.9219	7	0.27559	0.7	17.78
13/32	0.40625	10.3187	8	0.31496	0.8	20.32
27/64	0.421875	10.7156	9	0.35433	0.9	22.86
7/16	0.4375	11.1125	10	0.39370	1	25.4
29/64	0.453125	11.5094	11	0.43307	2	50.8
15/32	0.46875	11.9062	12	0.47244	3	76.2
31/64	0.484375	12.3031	13	0.51181	4	101.6
1/2	0.5	12.7000	14	0.55118	5	127.0
33/64	0.515625	13.0969	15	0.59055	6	152.4
17/32	0.53125	13.4937	16	0.62992	7	177.8
35/64	0.546875	13.8906	17	0.66929	8	203.2
9/16	0.5625	14.2875	18	0.70866	9	228.6
37/64	0.578125	14.6844	19	0.74803	10	254.0
19/32	0.59375	15.0812	20	0.78740	11	279.4
39/64	0.609375	15.4781	21	0.82677	12	304.8
5/8	0.625	15.8750	22	0.86614	13	330.2
41/64	0.640625	16.2719	23	0.90551	14	355.6
21/32	0.65625	16.6687	24	0.94488	15	381.0
43/64	0.671875	17.0656	25	0.98425	16	406.4
11/16	0.6875	17.4625	26	1.02362	17	431.8
45/64	0.703125	17.8594	27	1.06299	18	457.2
23/32	0.71875	18.2562	28	1.10236	19	482.6
47/64	0.734375	18.6531	29	1.14173	20	508.0
3/4	0.75	19.0500	30	1.18110	21	533.4
49/64	0.765625	19.4469	31	1.22047	22	558.8
25/32	0.78125	19.8437	32	1.25984	23	584.2
51/64	0.796875	20.2406	33	1.29921	24	609.6
13/16	0.8125	20.6375	34	1.33858	25	635.0
53/64	0.828125	21.0344	35	1.37795	26	660.4
27/32	0.84375	21.4312	36	1.41732	27	685.8
55/64	0.859375	21.8281	37	1.4567	28	711.2
7/8	0.875	22.2250	38	1.4961	29	736.6
57/64	0.890625	22.6219	39	1.5354	30	762.0
29/32	0.90625	23.0187	40	1.5748	31	787.4
59/64	0.921875	23.4156	41	1.6142	32	812.8
15/16	0.9375	23.8125	42	1.6535	33	838.2
61/64	0.953125	24.2094	43	1.6929	34	863.6
31/32	0.96875	24.6062	44	1.7323	35	889.0
63/64	0.984375	25.0031	45	1.7717	36	914.4

Metric Conversion Tables

1 Imperial gallon = 8 Imp pints = 1.16 US gallons = 277.42 cu in = 4.5459 litres

1 US gallon = 4 US quarts = 0.862 Imp gallon = 231 cu in = 3.785 litres

1 Litre = 0.2199 Imp gallon = 0.2642 US gallon = 61.0253 cu in = 1000 cc

Miles to Kilometres		Kilometres to Miles	
1	1.61	1	0.62
2	3.22	2	1.24
3	4.83	3	1.86
4	6.44	4	2.49
5	8.05	5	3.11
6	9.66	6	3.73
7	11.27	7	4.35
8	12.88	8	4.97
9	14.48	9	5.59
10	16.09	10	6.21
20	32.19	20	12.43
30	48.28	30	18.64
40	64.37	40	24.85
50	80.47	50	31.07
60	96.56	60	37.28
70	112.65	70	43.50
80	128.75	80	49.71
90	144.84	90	55.92
100	160.93	100	62.14

lb f ft to Kg f m		Kg f m to lb f ft		lb f/in^2: Kg f/cm^2		Kg f/cm^2: lb f/in^2	
1	0.138	1	7.233	1	0.07	1	14.22
2	0.276	2	14.466	2	0.14	2	28.50
3	0.414	3	21.699	3	0.21	3	42.67
4	0.553	4	28.932	4	0.28	4	56.89
5	0.691	5	36.165	5	0.35	5	71.12
6	0.829	6	43.398	6	0.42	6	85.34
7	0.967	7	50.631	7	0.49	7	99.56
8	1.106	8	57.864	8	0.56	8	113.79
9	1.244	9	65.097	9	0.63	9	128.00
10	1.382	10	72.330	10	0.70	10	142.23
20	2.765	20	144.660	20	1.41	20	284.47
30	4.147	30	216.990	30	2.11	30	426.70

Spanner size equivalents

AF		Whit	Fits	Metric Equivalent	Metric size A/F *	Inch Equivalent A/F *
4BA	0.248		9/64	6.3	7	0.276
2BA	0.32		3/16	8.1	8	0.315
					9	0.35
					10	0.39
7/16	0.44		1/4 UNF	11.2	11	0.413
	0.45	3/16	1/4 BSF	11.4	12	0.47
1/2	0.50		5/16 UNF	12.7	13	0.51
	0.53	1/4	5/16 BSF	13.5		
9/16	0.56		3/8 UNF	14.2	14	0.55
	0.604	5/16	3/8 BSF	15.3	15	0.59
5/8	0.63		7/16 Bolt	16	16	0.63
					17	0.67
11/16	0.69		7/16 Some nuts	17.5		
	0.72	3/8	7/16 BSF	18.3	18	0.71
3/4	0.76		1/2 UNF	19.3	19	0.75
					20	0.79
13/16	0.82			20.8		
	0.83	7/16	1/2 BSF	21.1	21	0.83
7/8	0.88		9/16 Some nuts	22.4	22	0.87
	0.93	1/2	9/16 BSF	23.6	23	0.91
15/16	0.94		5/8 UNF	23.8	24	0.945
					25	0.985
1"	1.01			25.6		
	1.02	9/16	5/8 BSF	25.9	26	1.02
1.1/16	1.07		5/8 Heavy UNF	27.2	27	1.06
	1.11	5/8	11/16 BSF	28.2	28	1.10
1.1/8	1.13		3/4 UNF	28.7	29	1.14
					30	1.18
	1.21	11/16	3/4 BSF	30.7	31	1.22
1.1/4	1.26		3/4 Heavy UNF	32.0	32	1.26
	1.31	3/4	7/8 BSF	33.3	33	1.3
1.5/16	1.32		7/8 UNF	33.5	34	1.34
					35	1.38
	1.49	7/8	1" BSF	37.8	36	1.42
					37	1.46

Safety first!

Professional motor mechanics are trained in safe working procedures. However enthusiastic you may be about getting on with the job in hand, do take the time to ensure that your safety is not put at risk. A moment's lack of attention can result in an accident, as can failure to observe certain elementary precautions.

There will always be new ways of having accidents, and the following points do not pretend to be a comprehensive list of all dangers; they are intended rather to make you aware of the risks and to encourage a safety-conscious approach to all work you carry out on your vehicle.

Essential DOs and DON'Ts

DON'T start the engine without first ascertaining that the transmission is in neutral.

DON'T suddenly remove the filler cap from a hot cooling system – cover it with a cloth and release the pressure gradually first, or you may get scalded by escaping coolant.

DON'T attempt to drain oil until you are sure it has cooled sufficiently to avoid scalding you.

DON'T grasp any part of the engine, exhaust or silencer without first ascertaining that it is sufficiently cool to avoid burning you.

DON'T allow brake fluid or antifreeze to contact the machine's paintwork or plastic components.

DON'T syphon toxic liquids such as fuel, brake fluid or antifreeze by mouth, or allow them to remain on your skin.

DON'T inhale dust – it may be injurious to health (see *Asbestos* heading).

DON'T allow any spilt oil or grease to remain on the floor – wipe it up straight away, before someone slips on it.

DON'T use ill-fitting spanners or other tools which may slip and cause injury.

DON'T attempt to lift a heavy component which may be beyond your capability – get assistance.

DON'T rush to finish a job, or take unverified short cuts.

DON'T allow children or animals in or around an unattended vehicle.

DON'T inflate a tyre to a pressure above the recommended maximum. Apart from overstressing the carcase and wheel rim, in extreme cases the tyre may blow off forcibly.

DO ensure that the machine is supported securely at all times. This is especially important when the machine is blocked up to aid wheel or fork removal.

DO take care when attempting to slacken a stubborn nut or bolt. It is generally better to pull on a spanner, rather than push, so that if slippage occurs you fall away from the machine rather than on to it.

DO wear eye protection when using power tools such as drill, sander, bench grinder etc.

DO use a barrier cream on your hands prior to undertaking dirty jobs – it will protect your skin from infection as well as making the dirt easier to remove afterwards; but make sure your hands aren't left slippery. Note that long-term contact with used engine oil can be a health hazard.

DO keep loose clothing (cuffs, tie etc) and long hair well out of the way of moving mechanical parts.

DO remove rings, wristwatch etc, before working on the vehicle – especially the electrical system.

DO keep your work area tidy – it is only too easy to fall over articles left lying around.

DO exercise caution when compressing springs for removal or installation. Ensure that the tension is applied and released in a controlled manner, using suitable tools which preclude the possibility of the spring escaping violently.

DO ensure that any lifting tackle used has a safe working load rating adequate for the job.

DO get someone to check periodically that all is well, when working alone on the vehicle.

DO carry out work in a logical sequence and check that everything is correctly assembled and tightened afterwards.

DO remember that your vehicle's safety affects that of yourself and others. If in doubt on any point, get specialist advice.

IF, in spite of following these precautions, you are unfortunate enough to injure yourself, seek medical attention as soon as possible.

Asbestos

Certain friction, insulating, sealing, and other products – such as brake linings, clutch linings, gaskets, etc – contain asbestos. *Extreme care must be taken to avoid inhalation of dust from such products since it is hazardous to health.* If in doubt, assume that they *do* contain asbestos.

Fire

Remember at all times that petrol (gasoline) is highly flammable. Never smoke, or have any kind of naked flame around, when working on the vehicle. But the risk does not end there – a spark caused by an electrical short-circuit, by two metal surfaces contacting each other, by careless use of tools, or even by static electricity built up in your body under certain conditions, can ignite petrol vapour, which in a confined space is highly explosive.

Always disconnect the battery earth (ground) terminal before working on any part of the fuel or electrical system, and never risk spilling fuel on to a hot engine or exhaust.

It is recommended that a fire extinguisher of a type suitable for fuel and electrical fires is kept handy in the garage or workplace at all times. Never try to extinguish a fuel or electrical fire with water.

Note: *Any reference to a 'torch' appearing in this manual should always be taken to mean a hand-held battery-operated electric lamp or flashlight. It does **not** mean a welding/gas torch or blowlamp.*

Fumes

Certain fumes are highly toxic and can quickly cause unconsciousness and even death if inhaled to any extent. Petrol (gasoline) vapour comes into this category, as do the vapours from certain solvents such as trichloroethylene. Any draining or pouring of such volatile fluids should be done in a well ventilated area.

When using cleaning fluids and solvents, read the instructions carefully. Never use materials from unmarked containers – they may give off poisonous vapours.

Never run the engine of a motor vehicle in an enclosed space such as a garage. Exhaust fumes contain carbon monoxide which is extremely poisonous; if you need to run the engine, always do so in the open air or at least have the rear of the vehicle outside the workplace.

The battery

Never cause a spark, or allow a naked light, near the vehicle's battery. It will normally be giving off a certain amount of hydrogen gas, which is highly explosive.

Always disconnect the battery earth (ground) terminal before working on the fuel or electrical systems.

If possible, loosen the filler plugs or cover when charging the battery from an external source. Do not charge at an excessive rate or the battery may burst.

Take care when topping up and when carrying the battery. The acid electrolyte, even when diluted, is very corrosive and should not be allowed to contact the eyes or skin.

If you ever need to prepare electrolyte yourself, always add the acid slowly to the water, and never the other way round. Protect against splashes by wearing rubber gloves and goggles.

Mains electricity and electrical equipment

When using an electric power tool, inspection light etc, always ensure that the appliance is correctly connected to its plug and that, where necessary, it is properly earthed (grounded). Do not use such appliances in damp conditions and, again, beware of creating a spark or applying excessive heat in the vicinity of fuel or fuel vapour. Also ensure that the appliances meet the relevant national safety standards.

Ignition HT voltage

A severe electric shock can result from touching certain parts of the ignition system, such as the HT leads, when the engine is running or being cranked, particularly if components are damp or the insulation is defective. Where an electronic ignition system is fitted, the HT voltage is much higher and could prove fatal.

English/American terminology

Because this book has been written in England, British English component names, phrases and spellings have been used throughout. American English usage is quite often different and whereas normally no confusion should occur, a list of equivalent terminology is given below.

English	American	English	American
Air filter	Air cleaner	Number plate	License plate
Alignment (headlamp)	Aim	Output or layshaft	Countershaft
Allen screw/key	Socket screw/wrench	Panniers	Side cases
Anticlockwise	Counterclockwise	Paraffin	Kerosene
Bottom/top gear	Low/high gear	Petrol	Gasoline
Bottom/top yoke	Bottom/top triple clamp	Petrol/fuel tank	Gas tank
Bush	Bushing	Pinking	Pinging
Carburettor	Carburetor	Rear suspension unit	Rear shock absorber
Catch	Latch	Rocker cover	Valve cover
Circlip	Snap ring	Selector	Shifter
Clutch drum	Clutch housing	Self-locking pliers	Vise-grips
Dip switch	Dimmer switch	Side or parking lamp	Parking or auxiliary light
Disulphide	Disulfide	Side or prop stand	Kick stand
Dynamo	DC generator	Silencer	Muffler
Earth	Ground	Spanner	Wrench
End float	End play	Split pin	Cotter pin
Engineer's blue	Machinist's dye	Stanchion	Tube
Exhaust pipe	Header	Sulphuric	Sulfuric
Fault diagnosis	Trouble shooting	Sump	Oil pan
Float chamber	Float bowl	Swinging arm	Swingarm
Footrest	Footpeg	Tab washer	Lock washer
Fuel/petrol tap	Petcock	Top box	Trunk
Gaiter	Boot	Torch	Flashlight
Gearbox	Transmission	Two/four stroke	Two/four cycle
Gearchange	Shift	Tyre	Tire
Gudgeon pin	Wrist/piston pin	Valve collar	Valve retainer
Indicator	Turn signal	Valve collets	Valve cotters
Inlet	Intake	Vice	Vise
Input shaft or mainshaft	Mainshaft	Wheel spindle	Axle
Kickstart	Kickstarter	White spirit	Stoddard solvent
Lower leg	Slider	Windscreen	Windshield
Mudguard	Fender		

Index

A
Air cleaner - 45

B
Ballast resistor - 55
Battery - 55
Brake pedal, rear - 73
Brakes:
    Adjustment - 82
    Fault diagnosis - 86
    Front - examination, dismantling & reassembly - 77
    Rear - examination, dismantling & reassembly - 79
    Specifications - 76
Bulb replacement - 57

C
Carburettor:
    Adjusting - 45
    Dismantling, inspection & reassembly - 42
    Fault diagnosis - 48
    Removal - 42
    Replacement - 44
    Specifications - 40
Centre stand - 71
Chain:
    Primary - 30, 32
    Final drive - 82
Clutch:
    Inspection - 29
    Fault diagnosis - 38, 39
    Reassembly & replacement - 32
    Removal & dismantling - 16
    Specifications - 10
Competition maintenance - 6
Condenser - 53
Contact breaker - 50, 51
Crankcases:
    Joining - 32
    Separating - 21
Crankshaft:
    Inspection - 27
    Removal - 22
    Replacement - 30
Cylinder barrel:
    Inspection - 27
    Replacement - 34
    Removal - 12
Cylinder head:
    Decarbonising - 29
    Inspection - 29
    Removal - 12
    Replacement - 34

E
Electrical system, specifications - 49
Engine:
    Dismantling - 12, 15, 16, 19, 21, 22, 26
    Fault diagnosis - 38
    Reassembly - 30, 32, 34
    Specifications - 10
Engine/gearbox unit:
    Removal - 11
    Replacement - 34
Exhaust system - 47

F
Final drive chain - 82
Final drive sprocket - 9, 34
Footrests - 73
Forks, front:
    Dismantling - 63
    Fault diagnosis - 75
    Inspection - 65, 67
    Reassembly & replacement - 68
    Removal - 59
Frame:
    Examination & renovation - 68
    Fault diagnosis - 75
Front brake - 77, 82
Front wheel:
    Examination - 76
    Replacement - 79
Front wheel bearings - 77
Fuel feed pipes - 41
Fuel system:
    Fault diagnosis - 48
    Specifications - 40
Fuel tank - 41
Fuel tap - 41

G
Gearbox:
    Components - examination - 29, 30
    Dismantling - 12, 15, 16, 19, 21, 22, 26
    Fault diagnosis - 39
    Reassembly - 30, 32, 34
    Specifications - 10
Gearbox/engine unit:
    Replacement - 34
    Removal - 11

H
Headlamp - 57
Horn - 57

Index

I

Ignition coil - 50
Ignition system:
    Fault diagnosis - 58
    Specifications - 49
Ignition timing - 51

K

Kickstarter - 16, 30

L

Lighting system - 55, 58
Lubrication system:
    Fault diagnosis - 48
    Specifications - 40

M

Magneto - 15, 34, 50
Main bearings - 26, 27
Maintenance, competition - 6
Maintenance, routine - 6

O

Oil seals - 27, 67
Ordering spare parts - 4

P

Petrol feed pipes - 41
Petrol/oil mix - 40
Petrol tank - 41
Petrol tap - 41
Piston:
    Inspection - 28
    Removal - 12
    Replacement - 34
Piston rings - 28
Primary chain - 30, 32
Prop stand - 71

R

Rear brake - 73, 79, 82
Rear suspension - 69, 71
Rear wheel:
    Bearings - 77, 79
    Examination & removal - 79
    Shock absorber - 82
    Sprocket - 82
Rectifier - 55
Resistor, ballast - 55
Routine maintenance - 6
Running-in - 38

S

Safety first - 90
Seat - 75
Small-end bearing - 28
Spare parts, ordering - 4
Spark plug - 64, 65
Speedometer - 57, 75
Sprockets:
    Final drive - 16, 34, 82
    Rear wheel - 82
Starting & running rebuilt engine - 38
Steering head - 63, 67
Stop lamp - 57
Stop lamp switch - 57
Suspension, fault diagnosis - 75
Suspension units, rear - 71
Swinging arm - 69
Switches - 57, 58

T

Tail lamp - 57
Tyres:
    Fault diagnosis - 86
    Pressures - 76
    Removal - 84
    Replacement - 84
    Security bolts - 86
    Sizes - 76
    Specifications - 76

W

Wheel bearings - 77, 79
Wheel, front - 76, 79, 86
Wheel, rear - 76, 79, 82, 86
Wiring - 58